Large Scale Interactive Fuzzy Multiobjective Programming

Studies in Fuzziness and Soft Computing

Editor-in-chief
Prof. Janusz Kacprzyk
Systems Research Institute
Polish Academy of Sciences
ul. Newelska 6
01-447 Warsaw, Poland
E-mail: kacprzyk@ibspan.waw.pl
http://www.springer.de/cgi-bin/search_book.pl?series=2941

Vol. 3. A. Geyer-Schulz
Fuzzy Rule-Based Expert Systems and Genetic Machine Learning, 2nd ed. 1996
ISBN 3-7908-0964-0

Vol. 4. T. Onisawa and J. Kacprzyk (Eds.)
Reliability and Safety Analyses under Fuzziness, 1995
ISBN 3-7908-0837-7

Vol. 5. P. Bosc and J. Kacprzyk (Eds.)
Fuzziness in Database Management Systems, 1995
ISBN 3-7908-0858-X

Vol. 6. E.S. Lee and Q. Zhu
Fuzzy and Evidence Reasoning, 1995
ISBN 3-7908-0880-6

Vol. 7. B.A. Juliano and W. Bandler
Tracing Chains-of-Thought, 1996
ISBN 3-7908-0922-5

Vol. 8. F. Herrera and J.L. Verdegay (Eds.)
Genetic Algorithms and Soft Computing, 1996
ISBN 3-7908-0956-X

Vol. 9. M. Sato et al.
Fuzzy Clustering Models and Applications, 1997
ISBN 3-7908-1026-6

Vol. 10. L.C. Jain (Ed.)
Soft Computing Techniques in Knowledge-based Intelligent Engineering Systems, 1997
ISBN 3-7908-1035-5

Vol. 11. W. Mielczarski (Ed.)
Fuzzy Logic Techniques in Power Systems, 1998
ISBN 3-7908-1044-4

Vol. 12. B. Bouchon-Meunier (Ed.)
Aggregation and Fusion of Imperfect Information, 1998
ISBN 3-7908-1048-7

Vol. 13. E. Orłowska (Ed.)
Incomplete Information: Rough Set Analysis, 1998
ISBN 3-7908-1049-5

Vol. 14. E. Hisdal
Logical Structures for Representation of Knowledge and Uncertainty, 1998
ISBN 3-7908-1056-8

Vol. 15. G.J. Klir and M.J. Wierman
Uncertainty-Based Information, 2nd ed., 1999
ISBN 3-7908-1242-0

Vol. 16. D. Driankov and R. Palm (Eds.)
Advances in Fuzzy Control, 1998
ISBN 3-7908-1090-8

Vol. 17. L. Reznik, V. Dimitrov and J. Kacprzyk (Eds.)
Fuzzy Systems Design, 1998
ISBN 3-7908-1118-1

Vol. 18. L. Polkowski and A. Skowron (Eds.)
Rough Sets in Knowledge Discovery 1, 1998
ISBN 3-7908-1119-X

Vol. 19. L. Polkowski and A. Skowron (Eds.)
Rough Sets in Knowledge Discovery 2, 1998
ISBN 3-7908-1120-3

Vol. 20. J.N. Mordeson and P.S. Nair
Fuzzy Mathematics, 1998
ISBN 3-7908-1121-1

Vol. 21. L.C. Jain and T. Fukuda (Eds.)
Soft Computing for Intelligent Robotic Systems, 1998
ISBN 3-7908-1147-5

Vol. 22. J. Cardoso and H. Camargo (Eds.)
Fuzziness in Petri Nets, 1999
ISBN 3-7908-1158-0

Vol. 23. P.S. Szczepaniak (Ed.)
Computational Intelligence and Applications, 1999
ISBN 3-7908-1161-0

Vol. 24. E. Orłowska (Ed.)
Logic at Work, 1999
ISBN 3-7908-1164-5

continued on page 218

Masatoshi Sakawa

Large Scale Interactive Fuzzy Multiobjective Programming

Decomposition Approaches

With 43 Figures
and 25 Tables

Physica-Verlag
A Springer-Verlag Company

Prof. Dr. Masatoshi Sakawa
Department of Industrial
and Systems Engineering
Faculty of Engineering
Hiroshima University
1-4-1 Kagamiyama
Higashi-Hiroshima 739-8527
Japan
E-mail: sakawa@msl.sys.hiroshima-u.ac.jp

ISSN 1434-9922

Cataloging-in-Publication Data applied for
Die Deutsche Bibliothek – CIP-Einheitsaufnahme
Sakawa, Masatoshi: Large scale interactive fuzzy multiobjective programming: decomposition approaches: with 25 tables / Masatoshi Sakawa. – Heidelberg; New York: Physica-Verl., 2000
(Studies in fuzziness and soft computing; Vol. 48)
ISBN 978-3-662-00386-2 ISBN 978-3-7908-1851-2 (eBook)
DOI 10.1007/978-3-7908-1851-2

Physica-Verlag is a company in the BertelsmannSpringer publishing group.

Softcover reprint of the hardcover 1st edition 2000

Hardcover Design: Erich Kirchner, Heidelberg

SPIN 10763587 88/2202-5 4 3 2 1 0 – Printed on acid-free paper

To my parents, Takeshige and Toshiko,
my wife Masako,
and my son Hideaki

Preface

The main characteristics of the real-world decision making problems facing humans today are large scale and have multiple objectives including economic, environmental, social, and technical issues.

Hence, actual decision making problems formulated as mathematical programming problems involve very large numbers of variables and constraints. Due to the high dimensionality of the problems, it becomes difficult to obtain optimal solutions for such large scale programming problems. Fortunately, however, most of the large scale programming problems arising in application almost always have a special structure that can be exploited. One familiar structure is the block angular structure to the constraints that can be used to formulate the subproblems for reducing both the processing time and memory requirements. With this observation, after the publication of the Dantzig-Wolfe decomposition method, both the dual and primal decomposition methods for solving large scale nonlinear programming problems with block angular structures have been proposed. Observe that the term *large scale programming problems* frequently means mathematical programming problems with block angular structures involving large numbers of variables and constraints.

Furthermore, it seems natural that the consideration of many objectives in the actual decision making process requires multiobjective approaches rather than a single objective. One of the major systems-analytic multiobjective approaches to decision making under constraints is multiobjective programming as a generalization of traditional single-objective programming. For such multiobjective programming problems, it is significant to realize that multiple objectives are often noncommensurable and conflict with each other. With this observation, in multiobjective programming problems, the notion of Pareto optimality or efficiency has been introduced instead of the optimality concept for single-objective problems. However, decisions with Pareto optimality or efficiency are not uniquely determined; the final decision must be selected by a decision maker, which well represents the subjective judgments, from the set of Pareto optimal or efficient solutions.

However, recalling the vagueness or fuzziness inherent in human judgments, two types of inaccuracies in human judgments should be incorporated in multiobjective programming problems. One is the experts' ambiguous understanding of the nature of the parameters in the problem-formulation process, and the other is the fuzzy goal of the decision maker for each of the objective functions. For handling and tackling such kinds of imprecisions or vaguenesses in human beings, it is not hard to imagine that the conventional multiobjective programming approaches, such as a deterministic or even a probabilistic approach, cannot be applied.

Naturally, simultaneous considerations of block angular structures, multiobjectiveness and fuzziness involved in the real-world decision making problems lead us to the new field of interactive multiobjective optimization for large scale programming problems under fuzziness. In this book, the author is concerned with introducing the latest advances in the new field of interactive multiobjective optimization for large scale linear and 0-1 programming problems under fuzziness on the basis of the author's continuing research. As further research directions, some of the most important related results, including interactive multiobjective optimization for linear fractional and nonlinear programming problems with block angular structures are also presented. Special stress is placed on interactive decision making aspects of fuzzy multiobjective optimization for human-centered systems in most realistic situations when dealing with fuzziness.

The intended readers of this book are senior undergraduate students, graduate students, researchers and practitioners in the fields of operations research, industrial engineering, management science, computer science, and other engineering disciplines that deal with the subjects of interactive multiobjective optimization for large scale programming problems under fuzziness. In order to master all the material discussed in this book, the readers would probably be required to have some background in linear algebra and mathematical programming. However, by skipping the mathematical details, much can be learned about large scale interactive fuzzy multiobjective programming for human-centered systems in most realistic settings without prior mathematical sophistication.

The author would like to express his sincere appreciation to Professor Janusz Kacprzyk of Polish Academy of Sciences, whose arrangements and warm encouragement made it possible for this book to be written. Special thanks should also be extended to Professor Yoshikazu Sawaragi, chairman of the Japan Institute of Systems Research and emeritus professor of Kyoto University, Department of Applied Mathematics and Physics, for his invariant stimulus and encouragement ever since the author's student days at Kyoto University. The author is also thankful to Dr. Masahiro Inuiguchi of Osaka University and Dr. Kazuya Sawada of Matsushita Electric Works, Ltd. for their contributions to Chapters 4 and 5, Dr. Kosuke Kato of Hiroshima University for his contribution to Chapters 7, 8 and 9, and Section 10.1, and Dr. Hitoshi Yano of Nagoya City University for his contribution to Section 10.2. Further thanks are due to Dr. Kosuke Kato of Hiroshima University for reviewing parts of the manuscript and for his helpful comments and suggestions. The author also wishes to thank all of his undergraduate and graduate students at Hiroshima University. Special thanks go to his former graduate students Ryuuji Mizouchi, Hideki Mohara, Keiichi Kubota and Toshihiro Ikegame of Hiroshima University for their invaluable assistance through discussions and computer works. Finally, the author would like to thank Dr. Martina Bihn, Physica-Verlag, Heidelberg, for her assistance in the publication of this book.

Hiroshima, December 1999 *Masatoshi Sakawa*

Contents

1. Introduction

1.1 Introduction and historical remarks

The increasing complexity of modern-day society has brought new problems involving very large numbers of variables and constraints. Due to the high dimensionality of the problems, it becomes difficult to obtain optimal solutions for such large scale programming problems. Fortunately, however, most of the large scale programming problems arising in application almost always have a special structure that can be exploited. One familiar structure is the block angular structure to the constraints that can be used to formulate the subproblems.

From such a point of view, in the early 1960s, Dantzig and Wolfe [13, 14] introduced the elegant and attractive decomposition method for linear programming problems. The Dantzig-Wolfe decomposition method, when applied to large scale linear programming problems with block angular structures, implies that the entire problem can be solved by solving a coordinated sequence of independent subproblems, and the process of coordination is shown to be finite.

After the publication of the Dantzig-Wolfe decomposition method, the subsequent works on large scale linear and nonlinear programming problems with block angular structures have been numerous [13, 14, 32, 34, 39, 40, 41, 42, 58, 69, 104, 123, 137]. Among the nonlinear extensions of the decomposition method, the dual decomposition method proposed by Lasdon [57] and the primal decomposition method proposed by Geoffrion [26] are well-known for solving large scale nonlinear programming problems with block angular structures. A brief and unified survey of major approaches to large scale mathematical programming proposed before 1970 can be found in the papers by Geoffrion [24, 25]. More comprehensive discussions of the major large scale mathematical programming proposed through the early 1970s can also be found in Lasdon [58] and Wismer [137].

Since the First International Conference on Multiple Criteria Decision Making, held at the University of South Carolina in 1972 [10], it has been increasingly recognized that most of the real-world decision making problems usually involve multiple, noncommensurable, and conflicting objectives which should be considered simultaneously. One of the major systems-analytic multiobjective approaches to decision making under constraints is multiobjective optimization as a generalization of traditional single-objective optimization. For such multiobjective optimization problems, it is significant to realize that multiple objectives are often noncommensurable and cannot be combined into a single objective. Moreover, the objectives usually conflict with each other and any improvement of one objective can be achieved only at the expense of another. With this observation, in multiobjective optimization, the notion of Pareto optimality or efficiency has been introduced instead of the optimality concept for single-objective optimization. However, decisions with Pareto optimality or efficiency are not uniquely determined; the final decision must be selected from among the set of Pareto optimal or efficient solutions. Consequently, the aim in solving multiobjective optimization problems is to derive a compromise or satisficing[†] solution of a decision maker which is also Pareto optimal based on subjective value judgments [8, 32, 78].

The interactive programming approaches, which assume that the decision maker is able to give some preference information on a local level to a particular solution, were first initiated by Geoffrion et al. [27] and further developed by many researchers such as Chankong and Haimes [8], Choo and Atkins [9], Oppenheimer [68], Sakawa [75], Sakawa and Seo [104], Sakawa and Yano [112], Steuer and Choo [130], and Wierzbicki [135].

However, considering the vague nature of the decision maker's judgments in multiobjective optimization problems, fuzzy programming approaches, first presented by Zimmermann [142] and further studied by Leberling [61], Hannan [33] and Sakawa [76] for multiobjective linear programming problems, seem to be very applicable and promising for solving multiobjective optimization problems.

In these fuzzy approaches, however, it has been implicitly assumed that the fuzzy decision or the minimum operator of Bellman and Zadeh is the

† "Satisficing" is a term or concept defined by March and Simon [65]. An alternative is satisficing if: (1) there exists a set of criteria that describes minimally satisfactory alternatives, and (2) the alternative in question meets or exceeds all these criteria.

proper representation of the decision maker's fuzzy preferences. Therefore, these approaches are preferable only when the decision maker feels that the fuzzy decision or the minimum-operator is appropriate when combining the fuzzy goals and/or constraints. However, such situations seem to rarely occur, and consequently it becomes evident that an interaction with the decision maker is necessary.

Assuming that the decision maker has a fuzzy goal for each of the objective functions in multiobjective programming problems, Sakawa et al. [78, 112] proposed several interactive fuzzy decision making methods by incorporating the desirable features of the interactive approaches into the fuzzy programming.

However, when formulating the multiobjective programming problem which closely describes and represents the real decision situation, various factors of the real system should be reflected in the description of the objective functions and the constraints. Naturally these objective functions and the constraints involve many parameters whose possible values may be assigned by the experts. In the traditional approaches, such parameters are fixed at some values in an experimental and/or subjective manner through the experts' understanding of the nature of the parameters.

In most practical situations, however, it is natural to consider that the possible values of these parameters are often only ambiguously known to the experts. In this case, it may be more appropriate to interpret the experts' understanding of the parameters as fuzzy numerical data which can be represented by means of fuzzy subsets of the real line known as fuzzy numbers by Dubois and Prade, [18, 19]. The resulting multiobjective programming problem involving fuzzy parameters would be viewed as the more realistic version of the conventional one.

In the mid 1980s, in order to deal with the multiobjective linear, linear fractional and nonlinear programming problems with fuzzy parameters characterized by fuzzy numbers, Sakawa and Yano [113, 114, 118] introduced the concept of α-multiobjective programming and (M-) α-Pareto optimality based on the α-level sets of the fuzzy numbers. They presented several interactive decision making methods not only in objective spaces but also in membership spaces to derive the satisficing solution for the decision maker efficiently from an (M-) α-Pareto optimal solution set for multiobjective linear, linear fractional and nonlinear programming problems as a generalization of their previous results [115, 116, 117, 119].

Under these circumstances, as a first attempt, Sakawa et al. [79, 80] considered large scale linear programming problems with block angular structures by incorporating both the fuzzy goal and fuzzy constraints of the decision maker. Having elicited the linear membership functions which well represent the fuzzy goal and fuzzy constraints, if the convex fuzzy decision was adopted for combining them, it was shown that, under some appropriate conditions, the formulated problem can be reduced to a number of independent linear subproblems and the satisficing solution for the decision maker is directly obtained just only by solving the subproblems.

In the framework of the fuzzy decision of Bellman and Zadeh [3], Sakawa et al. [81, 103], also considered fuzzy programming approaches to large scale multiobjective linear programming problems with block angular structures through the Dantzig-Wolfe decomposition method [13, 14]. By extending the framework of the fuzzy decision of Bellman and Zadeh, Sakawa et al. [102] proposed an interactive fuzzy satisficing method through the combination of the desirable features of both the interactive fuzzy satisficing methods [78] and the Dantzig-Wolfe decomposition method [13, 14]. Furthermore, in contrast to the large scale multiobjective linear programming problems with block angular structures, by considering the experts' vague or fuzzy understanding of the nature of the parameters in the problem-formulation process, Sakawa et al. formulated large scale multiobjective linear programming problems with block angular structures involving fuzzy numbers. Using the (M-) α-Pareto optimality concepts, Sakawa et al. [84, 85, 86, 87, 91, 93] presented several interactive decision making methods, which utilize the Dantzig-Wolfe decomposition method, not only in objective function spaces but also in membership function spaces to derive the satisficing solution for the decision maker efficiently from an (M-) α-Pareto optimal solution set. These results were immediately extended to deal with large scale multiobjective linear fractional programming problems with block angular structures involving fuzzy numbers [88, 89, 90].

So far, we have restricted ourselves to mathematical programming problems with continuous variables. The importance of mathematical programming problems with discrete variables, called integer programming problems or discrete optimization problems is well established, and, for some particular problems, elegant solution procedures have been achieved. Unfortunately, however, although the branch and bound strategy is known to be the most powerful and general-purpose solution method, there exist no exact algorithms for practical real-world problems. Hence, some powerful approximate algorithms are required for solving general discrete programming problems arising in real-world applications.

Genetic algorithms [43], initiated by Holland, his colleagues and his students at the University of Michigan in the 1970s, as stochastic search techniques based on the mechanism of natural selection and natural genetics, have received a great deal of attention regarding their potential as optimization techniques for solving discrete optimization problems or other hard optimization problems. Although genetic algorithms were not much known at the beginning, after the publication of Goldberg's book [28], genetic algorithms have recently attracted considerable attention in a number of fields as a methodology for optimization, adaptation and learning. As we look at recent applications of genetic algorithms to optimization problems, especially to various kinds of single-objective discrete optimization problems and/or to other hard optimization problems, we can see continuing advances [2, 16, 23, 67, 110].

As a natural extension of single-objective 0-1 programming problems, Sakawa et al. [82, 98] formulated multiobjective multidimensional 0-1 knapsack problems by assuming that the decision maker may have a fuzzy goal for each of the objective functions. After eliciting the linear membership functions, the fuzzy decision of Bellman and Zadeh [3] was adopted for combining these functions. For deriving a satisficing solution for the decision maker by solving the formulated problem, a genetic algorithm with double strings [82, 98], which generates only feasible solutions without using penalty functions for treating the constraints, was proposed. Also, through the combination of the desirable features of both the interactive fuzzy satisficing methods for continuous variables [78] and the genetic algorithm with double strings [96], an interactive fuzzy satisficing method to derive a satisficing solution for the decision maker to multiobjective multidimensional 0-1 knapsack problems was proposed [96, 106, 107]. Furthermore, they extended the proposed method to deal with multiobjective multidimensional 0-1 knapsack problems involving fuzzy numbers [108, 109].

For dealing with multiobjective multidimensional 0-1 knapsack problems with block angular structures, using triple string representation, Kato and Sakawa [46] presented a genetic algorithm with decomposition procedures. By incorporating the fuzzy goals of the decision maker, Kato and Sakawa [47] also proposed both fuzzy programming and interactive fuzzy programming for multiobjective multidimensional 0-1 knapsack problems with block angular structures [47]. Furthermore, Kato and Sakawa formulated large scale multiobjective multidimensional 0-1 knapsack problems with block angular structures involving fuzzy numbers. Using the (M-) α-Pareto optimality concepts, they [49, 50, 51, 52] presented several interactive decision making methods through genetic algorithms with decomposition procedures.

1.2 Organization of the book

The organization of each chapter is briefly summarized as follows.

Chapter 2 is devoted to mathematical preliminaries, which will be used throughout the remainder of this book. Starting with several basic definitions involving fuzzy sets, operations on fuzzy sets, especially fuzzy numbers, are outlined. Bellman and Zadeh's approach to decision making in a fuzzy environment, called fuzzy decision, is then examined. Fundamental notions and methods of multiobjective, interactive multiobjective, and interactive fuzzy multiobjective linear programming are briefly reviewed. A brief discussion of genetic algorithms is also given.

In Chapter 3, the Dantzig-Wolfe decomposition method for large scale linear programming problems with block angular structures is explained. The basic procedure and some of its variants are also introduced.

Chapter 4 treats large scale linear programming problems with block angular structures. Considering the vague or fuzzy nature of human judgments, both the fuzzy goal and fuzzy constrains of the decision maker are introduced. Having determined the corresponding membership functions, following the convex fuzzy decision for combining them, under suitable conditions, it is shown that the formulated problem can be reduced to a number of independent linear subproblems and the satisficing solution for the decision maker is directly obtained just only by solving the subproblems.

Chapter 5 can be viewed as the multiobjective version of Chapter 4 and is mainly concerned with interactive fuzzy multiobjective linear programming as well as fuzzy multiobjective linear programming.

In Chapter 6, in contrast to the multiobjective linear programming problems with block angular structures discussed thus far, by considering the experts' vague or fuzzy understanding of the nature of the parameters in the problem-formulation process, multiobjective linear programming problems with block angular structures involving fuzzy parameters are formulated. Through the introduction of extended Pareto optimality concepts, interactive multiobjective programming using the Dantzig-Wolfe decomposition method, both without and with the fuzzy goals of the decision maker, for deriving a satisficing solution for the decision maker from the extended Pareto optimal solution set are presented together with detailed numerical examples and experiments.

Chapter 7 presents a detailed treatment of genetic algorithms with decomposition procedures as developed for large scale 0-1 knapsack problems with block angular structures. Through the introduction of a triple string representation and the corresponding decoding algorithm, it is shown that a potential solution satisfying not only block constraints but also coupling constraints can be obtained for each individual. The chapter also includes several numerical experiments.

Chapter 8 can be viewed as the multiobjective version of Chapter 7 and treats large scale multiobjective 0-1 knapsack problems with block angular structures by incorporating the fuzzy goals of the decision maker. On the basis of the genetic algorithm with decomposition procedures presented in Chapter 7, interactive fuzzy multiobjective 0-1 programming as well as fuzzy multiobjective 0-1 programming are introduced together with several numerical experiments.

In Chapter 9, as the 0-1 version of Chapter 6, large scale multiobjective 0-1 knapsack problems with block angular structures involving fuzzy parameters are formulated. Along the same line as in Chapter 6, through the introduction of extended Pareto optimality concepts, interactive decision making methods using the genetic algorithms with decomposition procedures, both without and with the fuzzy goals of the decision maker, for deriving a satisficing solution efficiently from an extended Pareto optimal solution set are presented. Several numerical experiments are also given.

Finally, Chapter 10 outlines related topics including multiobjective linear fractional and nonlinear programming problems with block angular structures.

2. Mathematical Preliminaries

This chapter is devoted to mathematical preliminaries, which will be used in the remainder of this book. Starting with several basic definitions involving fuzzy sets, operations on fuzzy sets, especially fuzzy numbers, are outlined. Bellman and Zadeh's approach to decision making in a fuzzy environment, called fuzzy decision, is then examined. Fundamental notions and methods of multiobjective, interactive multiobjective, and interactive fuzzy multiobjective linear programming are briefly reviewed. A brief discussion of genetic algorithms is also given.

2.1 Fuzzy sets

In general, a fuzzy set initiated by Zadeh [139] is defined as follows:

Definition 2.1.1 (Fuzzy sets).

Let X denote a universal set. Then a fuzzy subset $\tilde{A}$ of X is defined by its membership function

$$\mu_{\tilde{A}} : X \to [0,1] \tag{2.1}$$

which assigns to each element $x \in X$ a real number $\mu_{\tilde{A}}(x)$ in the interval $[0,1]$, where the value of $\mu_{\tilde{A}}(x)$ at x represents the grade of membership of x in $\tilde{A}$. Thus, the nearer the value of $\mu_{\tilde{A}}(x)$ is unity, the higher the grade of membership of x in $\tilde{A}$.

A fuzzy subset $\tilde{A}$ can be characterized as a set of ordered pairs of element x and grade $\mu_{\tilde{A}}(x)$ and is often written

$$\tilde{A} = \{(x, \mu_{\tilde{A}}(x)) \mid x \in X\}. \tag{2.2}$$

When the membership function $\mu_{\tilde{A}}(x)$ contains only the two points 0 and 1, then $\mu_{\tilde{A}}(x)$ is identical to the characteristic function $c_A : X \to \{0,1\}$, and hence, $\tilde{A}$ is no longer a fuzzy subset, but an ordinary set A.

As is well known, an ordinary set A is expressed as

$$A = \{x \in X \mid c_A(x) = 1\}, \tag{2.3}$$

through its characteristic function

$$c_A(x) = \begin{cases} 1, & x \in A \\ 0, & x \notin A. \end{cases} \tag{2.4}$$

Figure 2.1 illustrates the membership function $\mu_{\tilde{A}}(x)$ of a fuzzy subset $\tilde{A}$ together with the characteristic function $c_A(x)$ of an ordinary set A.

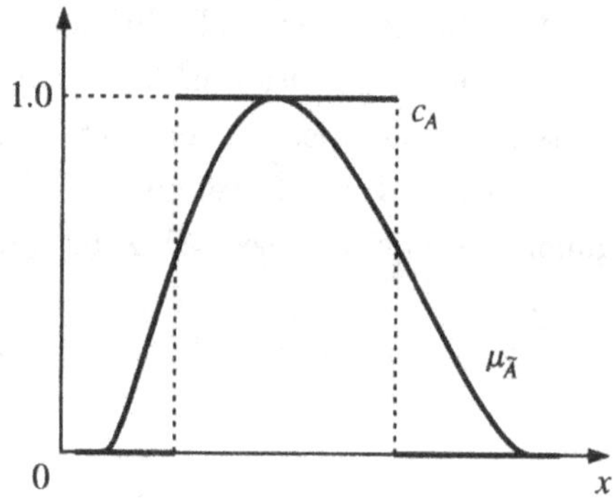

Figure 2.1. Membership function and characteristic function.

Observe that the membership function is an obvious extension of the idea of a characteristic function of an ordinary set because it takes on values between 0 and 1, not only 0 and 1.

As can be easily understood from the definition, a fuzzy subset is always defined as a subset of a universal set X. For the sake of convenience, a fuzzy subset is usually called a fuzzy set by omitting the term "sub." To distinguish an ordinary set from a fuzzy set, an ordinary set is called a nonfuzzy set or a crisp set. A fuzzy set is often denoted by $\tilde{A}, \tilde{B}, \tilde{C}, \ldots$, but it is sometimes written as $A, B, C, \ldots$, for simplicity in the notation.

The following basic notions are defined for fuzzy sets.

(1) ***Support***: The support of a fuzzy set A on X, denoted by supp (A), is the set of points in X at which $\mu_A(x)$ is positive, i.e.,

$$\text{supp}\,(A) = \{x \in X \mid \mu_A(x) > 0\}. \tag{2.5}$$

(2) *Height*: The height of a fuzzy set A on X, denoted by hgt(A), is the least upper bound of $\mu_A(x)$, i.e.,

$$\text{hgt}\,(A) = \sup_{x \in X} \mu_A(x). \tag{2.6}$$

(3) *Normal*: A fuzzy set A on X is said to be normal if its height is unity, i.e., if there is $x \in X$ such that $\mu_A(x) = 1$. If it is not normal, a fuzzy set is said to be subnormal.

(4) *Empty*: A fuzzy set A on X is empty, denoted by $\emptyset$, if and only if $\mu_A(x) = 0$ for all $x \in X$. Obviously, the universal set X can be viewed as a fuzzy set whose membership function is $\mu_X(x) = 1$ for all $x \in X$.

Observe that a nonempty subnormal fuzzy set A can be normalized by dividing $\mu_A(x)$ by its hgt(A).

Several set-theoretic operations involving fuzzy sets originally proposed by Zadeh [139] are as follows:

(1) *Equality*: The fuzzy sets A and B on X are equal, denoted by $A = B$, if and only if their membership functions are equal everywhere on X:

$$A = B \Leftrightarrow \mu_A(x) = \mu_B(x) \quad \text{for all} \quad x \in X. \tag{2.7}$$

(2) *Containment*: The fuzzy set A is contained in B (or a subset of B), denoted by $A \subseteq B$, if and only if their membership function is less or equal to that of B everywhere on X:

$$A \subseteq B \Leftrightarrow \mu_A(x) \leq \mu_B(x) \quad \text{for all} \quad x \in X. \tag{2.8}$$

(3) *Complementation*: The complement of a fuzzy set A on X, denoted by $\bar{A}$, is defined by

$$\mu_{\bar{A}}(x) = 1 - \mu_A(x) \quad \text{for all} \quad x \in X. \tag{2.9}$$

(4) *Intersection*: The intersection of two fuzzy sets A and B on X, denoted by $A \cap B$, is defined by

$$\mu_{A \cap B}(x) = \min\{\mu_A(x), \mu_B(x)\} \quad \text{for all} \quad x \in X. \tag{2.10}$$

(5) *Union*: The Union of two fuzzy sets A and B on X, denoted by $A \cup B$, is defined by

$$\mu_{A \cup B}(x) = \max\{\mu_A(x), \mu_B(x)\} \quad \text{for all} \quad x \in X. \tag{2.11}$$

Observe that the intersection $A \cap B$ is the largest fuzzy set which is contained in both A and B and the union $A \cup B$ is the smallest fuzzy set containing both A and B since any fuzzy set C such that $C \subseteq A$, $C \subseteq B$ satisfies $C \subseteq A \cap B$ and any fuzzy set D such that $D \supseteq A$, $D \supseteq B$ satisfies $D \supseteq A \cup B$.

The intersection and the union of two fuzzy sets A and B, and the complement of a fuzzy set A are illustrated in Figure 2.2, from which it can be easily understood that these set-theoretic operations for fuzzy sets can be viewed as a natural extension of those for ordinary sets.

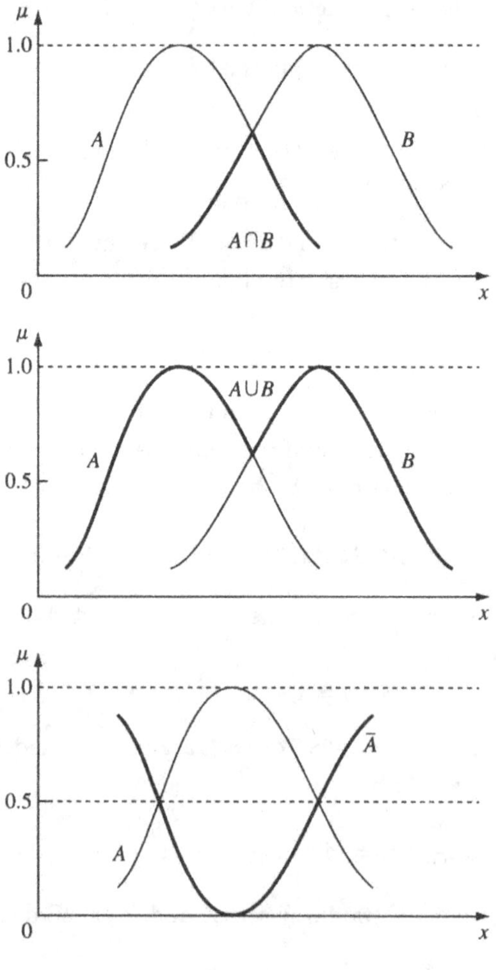

Figure 2.2. Set-theoretic operations for fuzzy sets.

The concept of α-level sets serves as an important transfer between ordinary sets and fuzzy sets. It also plays an important role in the construction of a fuzzy set by a series of ordinary sets.

Definition 2.1.2 (α-level set).

The α-level set of a fuzzy set A is defined as an ordinary set A_α for which the degree of its membership function exceeds the level α:

$$A_\alpha = \{x \mid \mu_A(x) \geq \alpha\}, \quad \alpha \in [0, 1]. \tag{2.12}$$

Observe that the α-level set A_α can be defined by the characteristic function

$$c_{A_\alpha} = \begin{cases} 1, & \mu_A(x) \geq \alpha, \\ 0, & \mu_A(x) < \alpha, \end{cases} \tag{2.13}$$

since it is an ordinary set. Actually, an α-level set is an ordinary set whose elements belong to the corresponding fuzzy set to a certain degree α.

It is clear that the following evident property holds for the α-level sets:

$$\alpha_1 \leq \alpha_2 \Leftrightarrow A_{\alpha_1} \supseteq A_{\alpha_2}. \tag{2.14}$$

This relationship is illustrated in Figure 2.3.

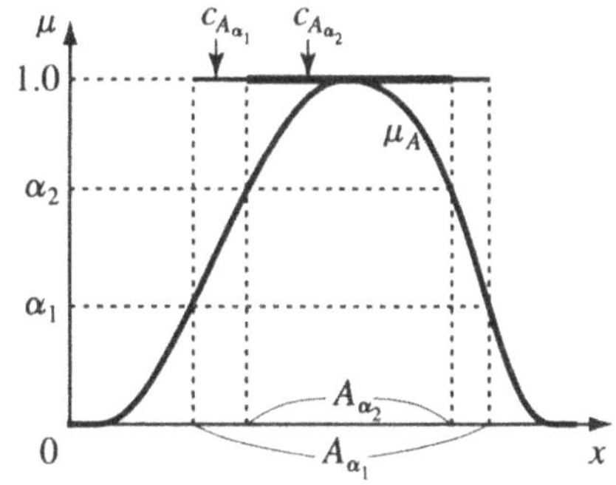

Figure 2.3. Examples of α-level sets.

From the definition of the α-level sets, it can be easily understood that the following basic properties hold:

$$(A \cup B)_\alpha = A_\alpha \cup B_\alpha, \tag{2.15}$$

$$(A \cap B)_\alpha = A_\alpha \cap B_\alpha. \tag{2.16}$$

2.2 Fuzzy numbers

Among fuzzy sets, numbers such as "approximately m" or "about n" can be defined as fuzzy sets of the real line R^1. Such fuzzy numbers are formally defined by Dubois and Prade [18, 19] as follows:

Definition 2.2.1 (Fuzzy numbers).
A fuzzy number $\tilde{m}$ is defined as any fuzzy set of the real line R^1, whose membership function $\mu_{\tilde{m}}(\cdot)$ is

(1) *A continuous mapping from R^1 to the closed interval* $[0, 1]$.
(2) $\mu_{\tilde{m}}(x) = 0$ *for all* $x \in (-\infty, a]$.
(3) *Strictly increasing and continuous on* $[a, m]$.
(4) $\mu_{\tilde{m}}(x) = 1$ *for* $x = m$.
(5) *Strictly decreasing and continuous on* $[m, b]$.
(6) $\mu_{\tilde{m}}(x) = 0$ *for all* $x \in [b, +\infty)$.

where a, b and m are real numbers, and $a < m < b$.

Figure 2.4 illustrates the graph of the possible shape of a fuzzy number $\tilde{m}$.

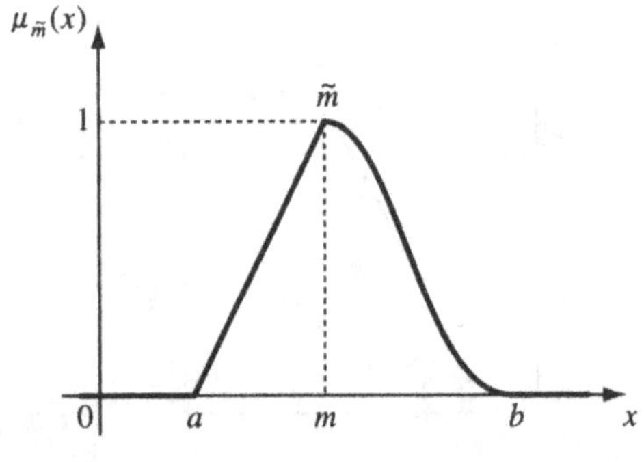

Figure 2.4. Fuzzy number.

Frequently, a fuzzy number $\tilde{m}$ is called positive (negative), denoted by $\tilde{m} > 0$ ($\tilde{m} < 0$), if its membership function $\mu_{\tilde{m}}(x)$ satisfies $\mu_{\tilde{m}}(x) = 0$, $\forall x < 0$ ($\forall x > 0$). A fuzzy number $\tilde{m}$ of triangular type is defined by a triangular membership function as

$$\mu_{\tilde{m}}(x) = \begin{cases} 0, & x < l, x > r \\ 1 - \dfrac{x-m}{l-m}, & l \leq x \leq m \\ 1 - \dfrac{x-m}{r-m}, & m \leq x \leq r. \end{cases} \tag{2.17}$$

Figure 2.5 shows an example of a membership function of a triangular fuzzy number $\tilde{m}$.

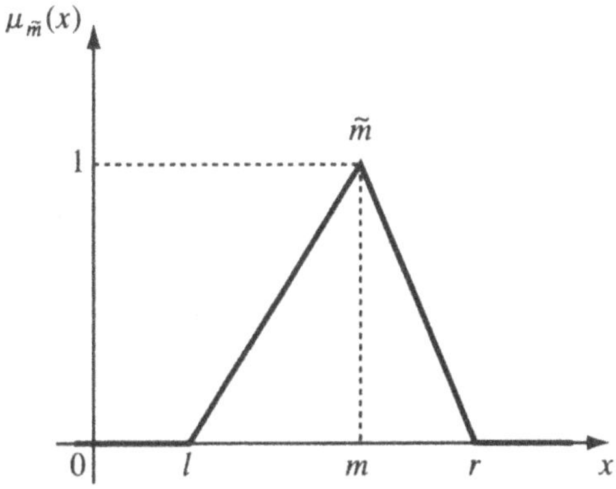

Figure 2.5. Triangular fuzzy number.

From the definition of a fuzzy number $\tilde{m}$, it is significant to note that the α-level set $\tilde{m}_\alpha$ of a fuzzy number $\tilde{m}$ can be represented by the closed interval which depends on the value of α as is shown in Figure 2.6. Namely,

$$\tilde{m}_\alpha = \{x \in R^1 \mid \mu_{\tilde{m}}(x) \geq \alpha\} = [m_\alpha^L, m_\alpha^R] \tag{2.18}$$

where m_α^L or m_α^R represents the left or right extreme point of the α-level set $\tilde{m}_\alpha$, respectively.

2.3 Fuzzy decision

Before discussing the original definitions of the fuzzy decision proposed by Bellman and Zadeh [3], it is quite significant to go back to their first explanation indicating the necessity for incorporating fuzzy sets into decision making processes. In their 1970 paper "Decision making in a fuzzy environment," Bellman and Zadeh [3] explain:

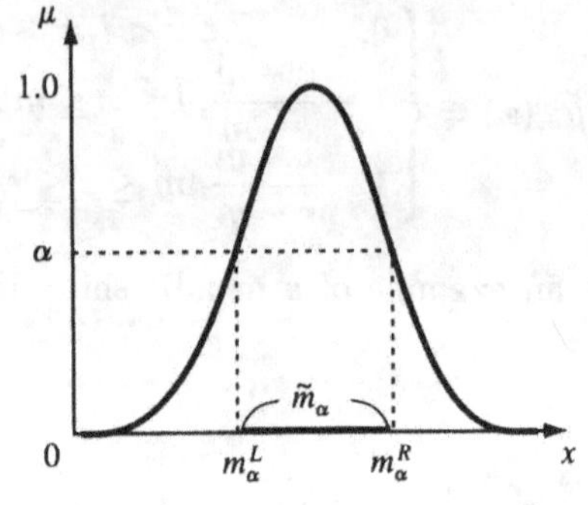

Figure 2.6. α-level set of fuzzy number $\tilde{m}$.

Much of the decision making in the real world takes place in an environment in which the goals, the constraints, and the consequences of possible actions are not known precisely. To deal quantitatively with imprecision, we usually employ the concepts and techniques of probability theory and, more particularly, the tools provided by decision theory, control theory, and information theory. In doing so, we are tacitly accepting the premise that imprecision – whatever its nature – can be equated with randomness. This, in our view, is a questionable assumption.

Specially, our contention is that there is a need for differentiation between *randomness* and *fuzziness*, with the later being a major source of imprecision in many decision processes. By fuzziness, we mean a type of imprecision which is associated with *fuzzy sets*, that is, classes in which there is no sharp transition from membership to nonmembership. For example, the class of *green objects* is a fuzzy set. So are the classes of objects characterized by such commonly used adjectives as large, small, substantial, significant, important, serious, simple, accurate, approximate, etc. Actually, in sharp contrast to the notion of a class or a set in mathematics, most of the classes in the real world do not have crisp boundaries which separate those objects which belong to a class from those which do not. In this connection, it is important to note that, in the discourse between humans, fuzzy statements such as "John is *several* inches taller than Jim," "x is *much larger* than y," "Corporation X has a *bright future*," "the stock market has suffered a *sharp decline*," convey information despite the imprecision of the meaning of the italicized word. In fact, it may be argued that the main distinction between human intelligence and machine intelligence lies in the ability of humans – an ability which present-day computers do not possess – to manipulate fuzzy concepts and respond to fuzzy instructions.

With this observation, Bellman and Zadeh [3] introduced three basic concepts: fuzzy goal, fuzzy constraint, and fuzzy decision and explored the application of these concepts to decision making processes under fuzziness.

Let us now introduce the conceptual framework for decision making in a fuzzy environment.

Let X be a given set of possible alternatives which contains the solution of a decision making problem under consideration.

A fuzzy goal G is a fuzzy set on X characterized by its membership function

$$\mu_G \;:\; X \to [0,1]. \tag{2.19}$$

A fuzzy constraint C is a fuzzy set on X characterized by its membership function

$$\mu_C \;:\; X \to [0,1]. \tag{2.20}$$

Realizing that both the fuzzy goal and fuzzy constraint are desired to be satisfied simultaneously, Bellman and Zadeh [3] defined the fuzzy decision D resulting from the fuzzy goal G and fuzzy constraint C as the intersection of G and C.

To be more explicit, the fuzzy decision of Bellman and Zadeh is the fuzzy set D on X defined as

$$D = G \cap C \tag{2.21}$$

and is characterized by its membership function

$$\mu_D(\boldsymbol{x}) = \min(\mu_G(\boldsymbol{x}), \mu_C(\boldsymbol{x})). \tag{2.22}$$

The maximizing decision is then defined as

$$\underset{\boldsymbol{x}\in X}{\text{maximize}}\, \mu_D(\boldsymbol{x}) = \underset{\boldsymbol{x}\in X}{\text{maximize}}\; \min(\mu_G(\boldsymbol{x}), \mu_C(\boldsymbol{x})). \tag{2.23}$$

More generally, the fuzzy decision D resulting from k fuzzy goals $G_1, \ldots, G_k$ and m fuzzy constraints $C_1, \ldots, C_m$ is defined by

$$D = G_1 \cap \cdots \cap G_k \cap C_1 \cap \cdots \cap C_m \tag{2.24}$$

and the corresponding maximizing decision is defined as

$$\begin{aligned} &\underset{\boldsymbol{x}\in X}{\text{maximize}}\, \mu_D(\boldsymbol{x}) \\ &= \underset{\boldsymbol{x}\in X}{\text{maximize}} \min(\mu_{G_1}(\boldsymbol{x}), \ldots, \mu_{G_k}(\boldsymbol{x}), \mu_{C_1}(\boldsymbol{x}), \ldots, \mu_{C_m}(\boldsymbol{x})). \end{aligned} \tag{2.25}$$

It is significant to realize here that in the fuzzy decision defined by Bellman and Zadeh [3], the fuzzy goals and the fuzzy constraints enter into the expression for D in exactly the same way. In other words, in the definition of the fuzzy decision, there is no longer a difference between the fuzzy goals and the fuzzy constraints.

However, depending on the situations, other aggregation patterns for the fuzzy goal G and the fuzzy constraint C may be worth considering. When fuzzy goals and fuzzy constraints have unequal importance, Bellman and Zadeh [3] also suggested the convex fuzzy decision defined by

$$\mu_D^{co}(\boldsymbol{x}) = \sum_{i=1}^{k} \alpha_i \mu_{G_i}(\boldsymbol{x}) + \sum_{j=1}^{m} \beta_j \mu_{C_j}(\boldsymbol{x}), \tag{2.26}$$

$$\sum_{i=1}^{k} \alpha_i + \sum_{j=1}^{m} \beta_j = 1, \quad \alpha_i, \beta_j \geq 0 \tag{2.27}$$

where the weighting coefficients reflect the relative importance among the fuzzy goals and constraints.

As an example of an alternative definition of a fuzzy decision, the product fuzzy decision defined by

$$\mu_D^{pr}(\boldsymbol{x}) = \left(\prod_{i=1}^{k} \mu_{G_i}(\boldsymbol{x}) \right) \cdot \left(\prod_{j=1}^{m} \mu_{C_j}(\boldsymbol{x}) \right) \tag{2.28}$$

has been proposed.

For the convex fuzzy decision or the product fuzzy decision, similar to the maximizing decision for the fuzzy decision, the maximizing decision to select $\boldsymbol{x}^*$ such that

$$\mu_D^{co}(\boldsymbol{x}^*) = \max_{\boldsymbol{x} \in X} \left[\sum_{i=1}^{k} \alpha_i \mu_{G_i}(\boldsymbol{x}) \right] \tag{2.29}$$

or

$$\mu_D^{pr}(\boldsymbol{x}^*) = \max_{\boldsymbol{x} \in X} \left[\left(\prod_{i=1}^{k} \mu_{G_i}(\boldsymbol{x}) \right) \cdot \left(\prod_{j=1}^{m} \mu_{C_j}(\boldsymbol{x}) \right) \right] \tag{2.30}$$

is also defined.

It should be noted here that among these three types of fuzzy decisions $\mu_D^{co}(\boldsymbol{x})$, $\mu_D^{pr}(\boldsymbol{x})$, and $\mu_D(\boldsymbol{x})$, the following relation holds:

$$\mu_D^{pr}(\boldsymbol{x}) \leq \mu_D(\boldsymbol{x}) \leq \mu_D^{co}(\boldsymbol{x}). \tag{2.31}$$

Example 2.3.1. Let $X = [0, \infty)$ be a set of alternatives. Suppose that we have a fuzzy goal G and a fuzzy constraint C expressed as "x should be much larger than 10" and "x should be substantially smaller than 30" where their membership functions are subjectively defined by

$$\mu_G(x) = \begin{cases} 0, & x \leq 10 \\ 1 - (1 + (0.1(x-10))^2)^{-1}, & x > 10, \end{cases}$$

$$\mu_C(x) = \begin{cases} 0, & x \geq 30 \\ (1 + x(x-30)^{-2})^{-1}, & x < 30. \end{cases}$$

The fuzzy decision, the convex fuzzy decision, and the product fuzzy decision for this situation are depicted in Figure 2.7.

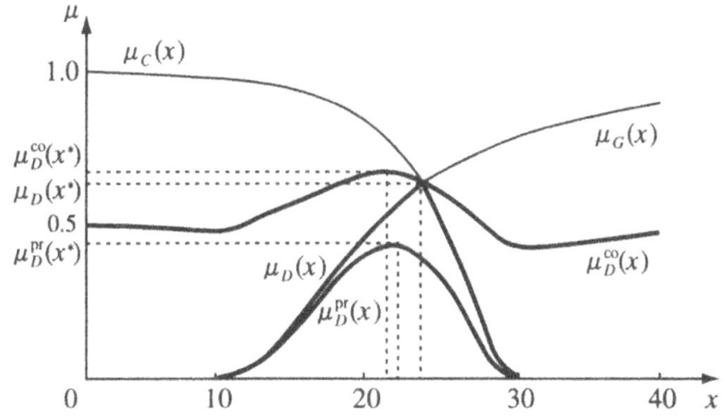

Figure 2.7. Fuzzy decision, convex fuzzy decision, and product fuzzy decision.

2.4 Multiobjective linear programming

The problem to optimize multiple conflicting linear objective functions simultaneously under given linear constraints is called the multiobjective linear programming problem and can be formulated as follows:

$$\left.\begin{array}{rl} \text{minimize} & z_1(\boldsymbol{x}) = \boldsymbol{c}_1\boldsymbol{x} \\ \text{minimize} & z_2(\boldsymbol{x}) = \boldsymbol{c}_2\boldsymbol{x} \\ & \vdots \\ \text{minimize} & z_k(\boldsymbol{x}) = \boldsymbol{c}_k\boldsymbol{x} \\ \text{subject to} & \boldsymbol{x} \in X \triangleq \{\boldsymbol{x} \in R^n \mid A\boldsymbol{x} \leq \boldsymbol{b},\ \boldsymbol{x} \geq \boldsymbol{0}\}, \end{array}\right\} \quad (2.32)$$

where

$$c_i = (c_{i1}, \ldots, c_{in}),\ i = 1, \ldots, k, \tag{2.33}$$
$$\boldsymbol{x} = (\boldsymbol{x}_1^T, \ldots, \boldsymbol{x}_n^T)^T, \tag{2.34}$$
$$A = \begin{bmatrix} a_{11} & \cdots & a_{1n} \\ \vdots & \ddots & \vdots \\ a_{m1} & \cdots & a_{mn} \end{bmatrix}, \tag{2.35}$$
$$\boldsymbol{b} = (b_1, \ldots, b_m)^T. \tag{2.36}$$

If we directly apply the notion of optimality for single-objective linear programming to this multiobjective linear programming, we arrive at the following notion of a complete optimal solution.

Definition 2.4.1 (Complete optimal solution).
$\boldsymbol{x}^*$ is said to be a complete optimal solution if and only if there exists $\boldsymbol{x}^* \in X$ such that $z_i(\boldsymbol{x}^*) \leq z_i(\boldsymbol{x})$, $i = 1, \ldots, k$, for all $\boldsymbol{x} \in X$.

However, in general, such a complete optimal solution that simultaneously minimizes all of the multiple objective functions does not always exist when the objective functions conflict with each other. Thus, instead of a complete optimal solution, a new solution concept, called Pareto optimality, is introduced in multiobjective linear programming [8, 78, 129, 140].

Definition 2.4.2 (Pareto optimal solution).
$\boldsymbol{x}^* \in X$ is said to be a Pareto optimal solution if and only if there does not exist another $\boldsymbol{x} \in X$ such that $z_i(\boldsymbol{x}) \leq z_i(\boldsymbol{x}^*)$ for all $i = 1, \ldots, k$, and $z_j(\boldsymbol{x}) < z_j(\boldsymbol{x}^*)$ for at least one $j \in \{1, \ldots, k\}$.

As can be seen from the definition, a Pareto optimal solution consists of an infinite number of points. A Pareto optimal solution is sometimes called a noninferior solution since it is not inferior to other feasible solutions.

In addition to Pareto optimality, the following weak Pareto optimality is defined as a slightly weaker solution concept than Pareto optimality.

Definition 2.4.3 (Weak Pareto optimal solution).
$\boldsymbol{x}^* \in X$ is said to be a weak Pareto optimal solution if and only if there does not exist another $\boldsymbol{x} \in X$ such that $z_i(\boldsymbol{x}) < z_i(\boldsymbol{x}^*)$, $1, \ldots, k$.

For notational convenience, let X^{CO}, X^P, or X^{WP} denote complete optimal, Pareto optimal, or weak Pareto optimal solution sets, respectively. Then from their definitions, it can be easily understood that the following relation holds:

$$X^{CO} \subseteq X^P \subseteq X^{WP}.$$

The details of multiobjective linear programming can be found in standard texts including Zeleny [140], Steuer [129], and Sakawa [78].

2.5 Interactive multiobjective linear programming

The STEP method (STEM) proposed by Benayoun et al. [4] seems to be known as one of the first interactive multiobjective linear programming techniques, but there have been some modifications and extensions (see, for example, Fichefet [22]; Choo and Atkins [9]). Essentially, the STEM algorithm consists of two major steps. Step 1 seeks a Pareto optimal solution that is near to the ideal point in the minimax sense. Step 2 requires the decision maker (DM) to compare the objective vector with the ideal vector and to indicate which objectives can be sacrificed, and by how much, in order to improve the current levels of unsatisfactory objectives. The STEM algorithm is quite simple to understand and implement, in the sense that the DM is required to give only the amounts to be sacrificed of some satisfactory objectives until all objectives become satisfactory. However, the DM will never arrive at the final solution if the DM is not willing to sacrifice any of the objectives. Moreover, in many practical situations, the DM will probably want to indicate directly the aspiration level for each objective rather than just specify the amount by which satisfactory objectives can be sacrificed.

Wierzbicki [135] developed a relatively practical interactive method called the reference point method (RPM) by introducing the concept of a reference point suggested by the DM which reflects in some sense the desired values of the objective functions. The basic idea behind the RPM is that the DM can specify reference values for the objective functions and change the reference objective levels interactively due to learning or improved understanding during the solution process. In this procedure, when the DM specifies a reference point, the corresponding scalarization problem is solved for generating the Pareto optimal solution which is, in a sense, close to the reference point or better than that if the reference point is attainable. Then the DM either chooses the current Pareto optimal solution or modifies the reference point to find a satisficing solution.

Since then, some similar interactive multiobjective programming methods have been developed along this line (see, for example, Steuer and Choo [130]). However, it is important to point out here that for dealing with the fuzzy goals of the DM for each of the objective functions of the multiobjective linear programming problem, Sakawa, Yano and Yumine [123] developed the extended fuzzy version of the RPM that supplies the DM with the trade-off information even if the fuzzy goals of the DM are not considered. Although the method will be outlined in the next section, it would certainly be appropriate to discuss here the RPM with trade-off information rather than the RPM proposed by Wierzbicki.

Consider the multiobjective linear programming problem expressed by (2.32). For each of the multiple conflicting objective functions $\boldsymbol{z}(\boldsymbol{x}) = (z_1(\boldsymbol{x}), \ldots, z_k(\boldsymbol{x}))^T$, assume that the DM can specify the so-called reference point $\bar{\boldsymbol{z}} = (\bar{z}_1, \ldots, \bar{z}_k)^T$ which reflects in some sense the desired values of the objective functions of the DM. Also assume that the DM can change the reference point interactively due to learning or improved understanding during the solution process. When the DM specifies the reference point $\bar{\boldsymbol{z}} = (\bar{z}_1, \ldots, \bar{z}_k)^T$, the corresponding Pareto optimal solution, which is, in the minimax sense, nearest to the reference point or better than that if the reference point is attainable, is obtained by solving the following minimax problem:

$$\left.\begin{array}{ll} \text{minimize} & \max\limits_{i=1,\ldots,k} \{z_i(\boldsymbol{x}) - \bar{z}_i\} \\ \text{subject to} & \boldsymbol{x} \in X, \end{array}\right\} \tag{2.37}$$

or equivalently

$$\left.\begin{array}{ll} \text{minimize} & v \\ \text{subject to} & z_i(\boldsymbol{x}) - \bar{z}_i \leq v, \ i = 1, \ldots, k \\ & \boldsymbol{x} \in X. \end{array}\right\} \tag{2.38}$$

The case of the two-objective functions in the z_1-z_2 plane is shown geometrically in Figure 2.8. For the two reference points $\bar{\boldsymbol{z}}^1 = (\bar{z}_1^1, \bar{z}_2^1)^T$ and $\bar{\boldsymbol{z}}^2 = (\bar{z}_1^2, \bar{z}_2^2)^T$ specified by the DM, solving the corresponding minimax problems yields the corresponding Pareto optimal solutions $\boldsymbol{z}^1(\boldsymbol{x}^1)$ and $\boldsymbol{z}^2(\boldsymbol{x}^2)$.

The relationships between the optimal solutions of the minimax problem and the Pareto optimal concept of the multiobjective linear programming can be characterized by the following two theorems.

Theorem 2.5.1. *If $\boldsymbol{x}^* \in X$ is a unique optimal solution of the minimax problem for any reference point $\bar{\boldsymbol{z}}$, then $\boldsymbol{x}^*$ is a Pareto optimal solution of the multiobjective linear programming problem.*

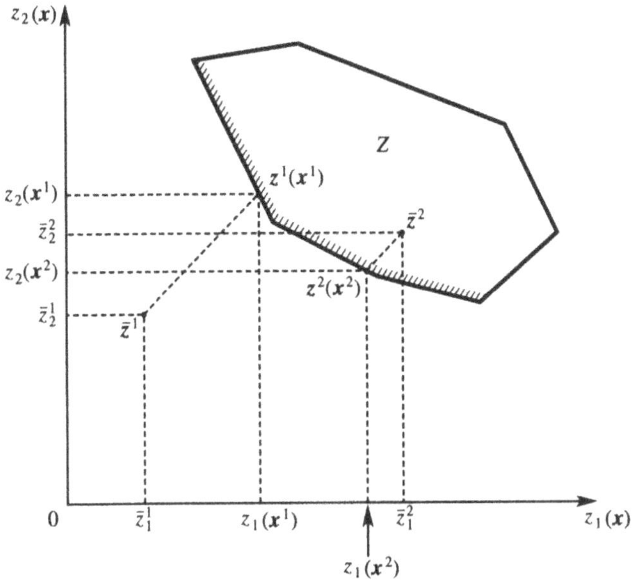

Figure 2.8. Graphical interpretation of minimax method.

It should be noted that only weak Pareto optimality is guaranteed if the uniqueness of a solution is not guaranteed.

Theorem 2.5.2. *If x^* is a Pareto optimal solution of the multiobjective linear programming problem, then x^* is an optimal solution of the minimax problem for some reference point $\bar{z}$.*

Now, given the Pareto optimal solution for the reference point specified by the DM by solving the corresponding minimax problem, the DM must either be satisfied with the current Pareto optimal solution or modify the reference point. To help the DM express a degree of preference, trade-off information between a standing objective function $z_1(x)$ and each of the other objective functions is very useful. Such a trade-off between $z_1(x)$ and $z_i(x)$ for each $i = 2, \ldots, k$ is easily obtainable since it is closely related to the strict positive simplex multipliers of the minimax problem. Let the simplex multipliers associated with the constraints of the minimax problem be denoted by π_i, $i = 1, \ldots, k$. If all $\pi_i > 0$ for each i, it can be proved that the following expression holds:

$$-\frac{\partial z_1(x)}{\partial z_i(x)} = \frac{\pi_1}{\pi_i}, \; i = 2, \ldots, k. \tag{2.39}$$

We can now construct the interactive algorithm to derive the satisficing solution for the DM from the Pareto optimal solution set. The steps marked with an asterisk involve interaction with the DM. Observe that this interactive multiobjective linear programming method can be interpreted as the reference point method (RPM) with trade-off information.

Interactive multiobjective linear programming

Step 0: Calculate the individual minimum $z_i^{\min} = \min_{\boldsymbol{x} \in X} z_i(\boldsymbol{x})$ and maximum $z_i^{\max} = \max_{\boldsymbol{x} \in X} z_i(\boldsymbol{x})$ of each objective function under the given constraints.

Step 1*: Ask the DM to select the initial reference point by considering the individual minimum and maximum. If the DM finds it difficult or impossible to identify such a point, ideal point $z_i^{\min} = \min_{\boldsymbol{x} \in X} z_i(\boldsymbol{x})$ can be used for that purpose.

Step 2: For the reference point specified by the DM, solve the corresponding minimax problem to obtain the Pareto optimal solution together with the trade-off rate information between the objective functions.

Step 3*: If the DM is satisfied with the current levels of the Pareto optimal solution, stop. Then the current Pareto optimal solution is the satisficing solution for the DM. Otherwise, ask the DM to update the current reference point by considering the current values of the objective functions together with the trade-off rates between the objective functions and return to Step 2.

It should be stressed to the DM that any improvement of one objective function can be achieved only at the expense of at least one of the other objective functions.

Further details of the theory, methods and applications of interactive multiobjective linear programming can be found in Steuer [129], and Sakawa [78].

2.6 Interactive fuzzy multiobjective linear programming

In the fuzzy approaches to multiobjective linear programming problems proposed by Zimmermann [142] and his successors [61, 33, 143], it has been implicitly assumed that the fuzzy decision of Bellman and Zadeh [3] is the proper representation of the fuzzy preferences of the decision maker (DM). Therefore, these approaches are preferable only when the DM feels that the fuzzy decision is appropriate when combining the fuzzy goals and/or constraints. However, such situations seem to occur rarely in practice and consequently it becomes evident that an interaction with the DM is necessary.

In this section, assuming that the DM has a fuzzy goal for each of the objective functions in multiobjective linear programming problems, we present an interactive fuzzy multiobjective linear programming method incorporating the desirable features of the interactive approaches into the fuzzy approaches.

Fundamental to the multiobjective linear programming is the concept of Pareto optimal solutions, also known as a noninferior solution.

However, considering the imprecise nature inherent in human judgments in multiobjective linear programming problems, the DM may have a fuzzy goal expressed as "$z_i(\boldsymbol{x})$ should be substantially less than or equal to some value p_i."

In a minimization problem, a fuzzy goal stated by the DM may be to achieve "substantially less than or equal to p_i." This type of statement can be quantified by eliciting a corresponding membership function.

To elicit a membership function $\mu_i(z_i(\boldsymbol{x}))$ from the DM for each of the objective functions $z_i(\boldsymbol{x})$, $i = 1, \ldots, k$, we first calculate the individual minimum $z_i^{\min} = \min_{\boldsymbol{x} \in X} \; z_i(\boldsymbol{x})$ and maximum $z_i^{\max} = \max_{\boldsymbol{x} \in X} \; z_i(\boldsymbol{x})$ of each objective function $z_i(\boldsymbol{x})$ under the given constraints.

Taking into account the calculated individual minimum and maximum of each objective function together with the rate of increase of membership of satisfaction, the DM must determine the subjective membership function $\mu_i(z_i(\boldsymbol{x}))$, which is a strictly monotone decreasing function with respect to $z_i(\boldsymbol{x})$. Here, it is assumed that $\mu_i(z_i(\boldsymbol{x})) = 0$ or $\to 0$ if $z_i(\boldsymbol{x}) \geq z_i^0$ and $\mu_i(z_i(\boldsymbol{x})) = 1$ or $\to 1$ if $z_i(\boldsymbol{x}) \leq z_i^1$.

So far, we have restricted ourselves to a minimization problem and consequently assumed that the DM has a fuzzy goal such as "$z_i(\boldsymbol{x})$ should be substantially less than or equal to p_i." In the fuzzy approaches, however, we can further treat a more general multiobjective linear programming problem in which the DM has two types of fuzzy goals expressed in words such as "$z_i(\boldsymbol{x})$ should be in the vicinity of r_i" (called fuzzy equal), "$z_i(\boldsymbol{x})$ should be substantially less than or equal to p_i" (called fuzzy min) or "$z_i(\boldsymbol{x})$ should be substantially greater than or equal to q_i" (called fuzzy max).

Such a generalized multiobjective linear programming problem may now be expressed as

$$\left.\begin{array}{lll} \text{fuzzy min} & z_i(\boldsymbol{x}) & i \in I_1 \\ \text{fuzzy max} & z_i(\boldsymbol{x}) & i \in I_2 \\ \text{fuzzy equal} & z_i(\boldsymbol{x}) & i \in I_3 \\ \text{subject to} & \boldsymbol{x} \in X & \end{array}\right\} \quad (2.40)$$

where $I_1 \cup I_2 \cup I_3 = \{1, \ldots, k\}$, $I_i \cap I_j = \emptyset$, $i, j = 1, 2, 3$, $i \neq j$.

Here "fuzzy min $z_i(\boldsymbol{x})$" or "fuzzy max $z_i(\boldsymbol{x})$" represents the fuzzy goal of the DM such as "$z_i(\boldsymbol{x})$ should be substantially less than or equal to p_i or greater than or equal to q_i," and "fuzzy equal $z_i(\boldsymbol{x})$" represents the fuzzy goal such as "$z_i(\boldsymbol{x})$ should be in the vicinity of r_i."

Concerning the membership function for the fuzzy goal of the DM such as "$z_i(\boldsymbol{x})$ should be in the vicinity of r_i," it is obvious that a strictly monotone increasing function $d_{iL}(z_i)$, $(i \in I_3)$ and a strictly monotone decreasing function $d_{iR}(z_i)$, $(i \in I_3)$, corresponding to the left and right sides of r_i must be determined through interaction with the DM.

Figures 2.9, 2.10 and 2.11 illustrate possible shapes of the fuzzy min, fuzzy max and fuzzy equal membership functions, respectively.

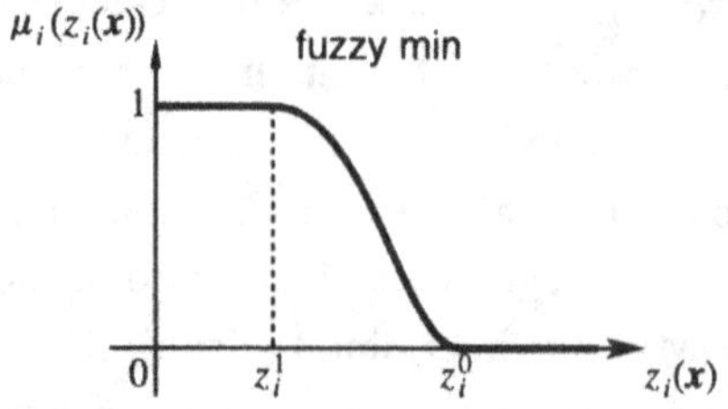

Figure 2.9. Fuzzy min membership function.

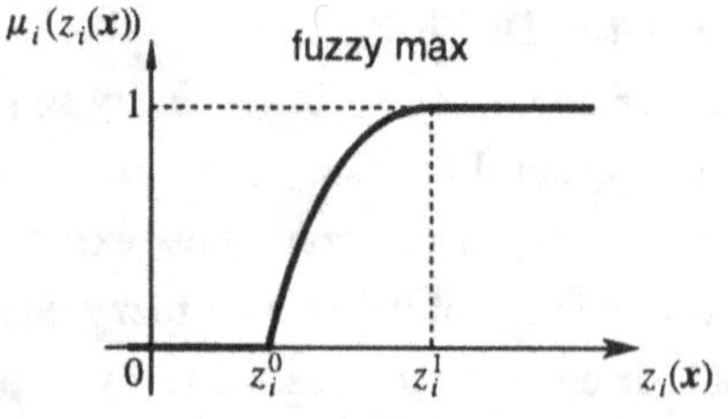

Figure 2.10. Fuzzy max membership function.

Having elicited the membership functions $\mu_i(z_i(\boldsymbol{x}))$, $i = 1, \ldots, k$, from the DM for each of the objective functions $z_i(\boldsymbol{x})$, $i = 1, \ldots, k$, the multi-

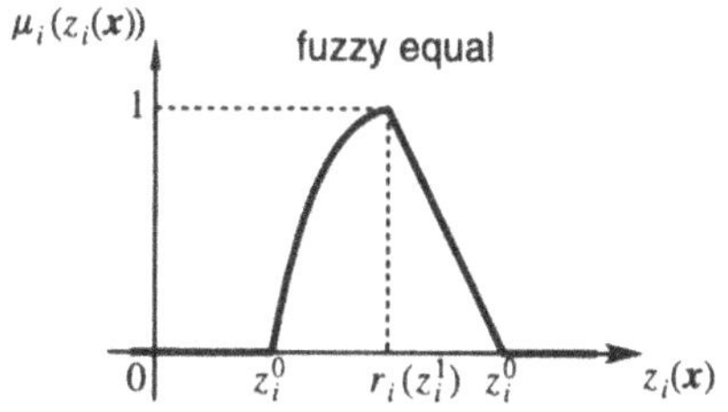

Figure 2.11. Fuzzy equal membership function.

objective linear programming problem and/or the generalized multiobjective linear programming problem can be converted into the fuzzy multiobjective optimization problem defined by

$$\underset{x \in X}{\text{maximize}} \ (\mu_1(z_1(x)), \mu_2(z_2(x)), \ldots, \mu_k(z_k(x))). \tag{2.41}$$

When the fuzzy equal is included in the fuzzy goals of the DM, it is desirable that $z_i(x)$ should be as close to r_i as possible. Consequently, the notion of Pareto optimal solutions defined in terms of objective functions cannot be applied. For this reason, we introduce the concept of M-Pareto optimal solutions which is defined in terms of membership functions instead of objective functions. M refers to membership.

Definition 2.6.1 (M-Pareto optimal solution).
$x^* \in X$ is said to be an M-Pareto optimal solution to the generalized multiobjective linear programming problem if and only if there does not exist another $x \in X$ such that $\mu_i(z_i(x)) \geq \mu_i(z_i(x^*))$ for all $i = 1, \ldots, k$, and $\mu_j(z_j(x)) > \mu_j(z_j(x^*))$ for at least one $j \in \{1, \ldots, k\}$.

By introducing a general aggregation function

$$\mu_D(\mu(z(x))) = \mu_D(\mu_1(z_1(x)), \mu_2(z_2(x)), \ldots, \mu_k(z_k(x))), \tag{2.42}$$

a general fuzzy multiobjective decision making problem can be defined by

$$\underset{x \in X}{\text{maximize}} \ \mu_D(\mu(z(x))). \tag{2.43}$$

Observe that the value of $\mu_D(\mu(z(x)))$ can be interpreted as representing an overall degree of satisfaction with the DM's multiple fuzzy goals.

Probably the most crucial problem in the fuzzy multiobjective decision making problem is the identification of an appropriate aggregation function which well represents the DM's fuzzy preferences. If $\mu_D(\cdot)$ can be explicitly identified, then the fuzzy multiobjective decision making problem reduces to a standard mathematical programming problem. However, this rarely happens, and as an alternative, an interaction with the DM is necessary for finding the satisficing solution of the fuzzy multiobjective decision making problem.

In the interactive fuzzy multiobjective linear programming method proposed by Sakawa, Yano and Yumine [123], after determining the membership functions $\boldsymbol{\mu}(\boldsymbol{z}(\boldsymbol{x})) = (\mu_1(z_1(\boldsymbol{x})), \ldots, \mu_k(z_k(\boldsymbol{x})))^T$ for each of the objective functions $\boldsymbol{z}(\boldsymbol{x}) = (z_1(\boldsymbol{x}), \ldots, z_k(\boldsymbol{x}))^T$, for generating a candidate for the satisficing solution which is also M-Pareto optimal, the DM is then asked to specify the aspiration levels of achievement for the membership values of all membership functions, called the reference membership levels. The reference membership levels can be viewed as natural extensions of the reference point of Wierzbicki [135] in objective function spaces.

For the DM's reference membership levels $\bar{\boldsymbol{\mu}} = (\bar{\mu}_1, \ldots, \bar{\mu}_k)^T$, the corresponding M-Pareto optimal solution, which is nearest to the requirements in the minimax sense or better than that if the reference membership levels are attainable, is obtained by solving the following minimax problem

$$\underset{\boldsymbol{x} \in X}{\text{minimize}} \max_{i=1,\ldots,k} \{\bar{\mu}_i - \mu_i(z_i(\boldsymbol{x}))\}, \tag{2.44}$$

or equivalently

$$\left.\begin{array}{ll} \text{minimize} & v \\ \text{subject to} & \bar{\mu}_i - \mu_i(z_i(\boldsymbol{x})) \leq v, \;\; i = 1, \ldots, k \\ & \boldsymbol{x} \in X. \end{array}\right\} \tag{2.45}$$

The relationships between the optimal solutions of the minimax problem and the M-Pareto optimal concept of the multiobjective linear programming problem can be characterized by the following theorem.

Theorem 2.6.1. *If $\boldsymbol{x}^* \in X$ is a unique optimal solution to the minimax problem for some $\bar{\mu}_i$, $i = 1, \ldots, k$, then $\boldsymbol{x}^*$ is an M-Pareto optimal solution to the generalized multiobjective linear programming problem.*

Theorem 2.6.2. *If $\boldsymbol{x}^*$ is an M-Pareto optimal solution to the generalized multiobjective linear programming problem with $0 < \mu_i(z_i(\boldsymbol{x}^*)) < 1$ holding for all i, then there exists $\bar{\mu}_i$, $i = 1, \ldots, k$, such that $\boldsymbol{x}^*$ is an optimal solution to the minimax problem.*

If all of the membership functions $\mu_i(z_i(\boldsymbol{x}))$, $i = 1, \ldots, k$, are linear, the minimax problem becomes a linear programming problem, and hence, we can obtain an optimal solution by directly applying the simplex method of linear programming.

However, with the strictly monotone decreasing or increasing membership functions, which may be nonlinear, the resulting minimax problem becomes a nonlinear programming problem. For notational convenience, denote the strictly monotone decreasing function for the fuzzy min and the right function of the fuzzy equal by $d_{iR}(z_i)$ $(i \in I_1 \cup I_3)$ and the strictly monotone increasing function for the fuzzy max and the left function of the fuzzy equal by $d_{iL}(z_i)$ $(i \in I_2 \cup I_3)$. Then in order to solve the formulated problem on the basis of the linear programming method, convert each constraint $\bar{\mu}_i - \mu_i(z_i(\boldsymbol{x})) \leq v$, $i = 1, \ldots, k$, of the minimax problem (2.45) into the following form using the strictly monotone property of $d_{iL}(\cdot)$ and $d_{iR}(\cdot)$:

$$\left.\begin{array}{ll} \text{minimize} & v \\ \text{subject to} & z_i(\boldsymbol{x}) \leq d_{iR}^{-1}(\bar{\mu}_i - v), \quad i \in I_1 \cup I_3 \\ & z_i(\boldsymbol{x}) \geq d_{iL}^{-1}(\bar{\mu}_i - v), \quad i \in I_2 \cup I_3 \\ & \boldsymbol{x} \in X. \end{array}\right\} \tag{2.46}$$

It is important to note here that, if the value of v is fixed, it can be reduced to a set of linear inequalities. Obtaining the optimal solution v^* to the above problem is equivalent to determining the minimum value of v so that there exists an admissible set satisfying the constraints of (2.46). Since v satisfies $\bar{\mu}_{\max} - 1 \leq v \leq \bar{\mu}_{\max}$, where $\bar{\mu}_{\max}$ denotes the maximum value of $\bar{\mu}_i$, $i = 1, \ldots, k$, we have the following method for solving this problem by combined use of the bisection method and the simplex method of linear programming. Here, when $\bar{\mu}_i - v \leq 0$, set $\bar{\mu}_i - v = 0$ in view of the constraints $\bar{\mu}_i - v \leq \mu_i(z_i(\boldsymbol{x}))$ for $0 \leq \mu_i(z_i(\boldsymbol{x})) \leq 1$, $i = 1, \ldots, k$.

Step 1: Set $v = \bar{\mu}_{\max}$ and test whether an admissible set satisfying the constraints of (2.46) exists or not using phase one of the simplex method. If an admissible set exists, proceed. Otherwise, the DM must reassess the membership function.

Step 2: Set $v = \bar{\mu}_{\max} - 1$ and test whether an admissible set satisfying the constraints of (2.46) exists or not using phase one of the simplex method. If an admissible set exists, set $v^* = \bar{\mu}_{\max} - 1$. Otherwise, go to the next step since the minimum v which satisfies the constraints of (2.46) exists between $\bar{\mu}_{\max} - 1$ and $\bar{\mu}_{\max}$.

Step 3: For the initial value of $v = \bar{\mu}_{\max} - 0.5$, update the value of v using the bisection method as follows:

$$\begin{cases} v_{n+1} = v_n - 1/2^{n+1} \text{ if an admissible set exists for } v_n, \\ v_{n+1} = v_n + 1/2^{n+1} \text{ if no admissible set exists for } v_n. \end{cases}$$

For each v_n, $n = 1, 2, \ldots$, test whether an admissible set of (2.46) exists or not using the sensitivity analysis technique for changes in the right-hand side of the simplex method and determine the minimum value of v satisfying the constraints of (2.46).

In this way, we can determine the optimal solution v^*. Then the DM selects an appropriate standing objective from among the objectives $z_i(\boldsymbol{x})$, $i = 1, \ldots, k$. For notational convenience in the following without loss of generality, let it be $z_1(\boldsymbol{x})$ and $1 \in I_1$. Then the following linear programming problem is solved for $v = v^*$:

$$\left.\begin{array}{rl} \text{minimize} & z_1(\boldsymbol{x}) \\ \text{subject to} & z_i(\boldsymbol{x}) \leq d_{iR}^{-1}(\bar{\mu}_i - v^*), \ \ i(\neq 1) \in I_1 \cup I_3 \\ & z_i(\boldsymbol{x}) \geq d_{iL}^{-1}(\bar{\mu}_i - v^*), \ \ i(\neq 1) \in I_2 \cup I_3 \\ & \boldsymbol{x} \in X. \end{array}\right\} \tag{2.47}$$

The DM must either be satisfied with the current M-Pareto optimal solution or act on this solution by updating the reference membership levels. In order to help the DM express a degree of preference, trade-off information between a standing membership function $\mu_1(z_1(\boldsymbol{x}))$ and each of the other membership functions is very useful. Such trade-off information is easily obtainable since it is closely related to the simplex multipliers of the problem (2.47).

Let the simplex multipliers corresponding to the constraints $z_i(\boldsymbol{x})$, $i = 2, \ldots, k$, of the linear problem (2.47) be denoted by $\pi_i^* = \pi_i(\boldsymbol{x}^*)$, $i = 2, \ldots, k$, where $\boldsymbol{x}^*$ is an optimal solution of (2.47). If $\boldsymbol{x}^*$ is a nondegenerate solution of (2.47) and all the constraints of (2.47) are active, then by using the results in Haimes and Chankong [31], the trade-off information between the objective functions can be represented by

$$-\frac{\partial z_1(\boldsymbol{x})}{\partial z_i(\boldsymbol{x})} = \pi_i^*, \ \ i = 2, \ldots, k. \tag{2.48}$$

Hence, by the chain rule, the trade-off information between the membership functions is given by

$$-\frac{\partial \mu_1(z_1(\boldsymbol{x}))}{\partial \mu_i(z_i(\boldsymbol{x}))} = -\frac{\partial \mu_1(z_1(\boldsymbol{x}))}{\partial z_1(\boldsymbol{x})} \frac{\partial z_1(\boldsymbol{x})}{\partial z_i(\boldsymbol{x})} \left\{ \frac{\partial \mu_i(z_i(\boldsymbol{x}))}{\partial z_i(\boldsymbol{x})} \right\}^{-1}, \ i = 2, \ldots, k. \tag{2.49}$$

Therefore, for each $i = 2, \ldots, k$, we have the following expression:

$$-\frac{\partial \mu_1(z_1(\boldsymbol{x}))}{\partial \mu_i(z_i(\boldsymbol{x}))} = \pi_i^* \frac{\partial \mu_1(z_1(\boldsymbol{x}))/\partial z_1(\boldsymbol{x})}{\partial \mu_i(z_i(\boldsymbol{x}))/\partial z_i(\boldsymbol{x})}, \quad i = 2, \ldots, k. \tag{2.50}$$

It should be stressed here that in order to obtain the trade-off rate information from (2.50), all the constraints of the problem (2.47), must be active. Therefore, if there are inactive constraints, it is necessary to replace $\bar{\mu}_i$ for inactive constraints by $\bar{\mu}_i(z_i(\boldsymbol{x}^*))$ and solve the corresponding problem to obtain the simplex multipliers.

We can now construct the interactive algorithm in order to derive the satisficing solution for the DM from the M-Pareto optimal solution set where the steps marked with an asterisk involve interaction with the DM. This interactive fuzzy multiobjective programming method can also be interpreted as the fuzzy version of the reference point method (RPM) with trade-off information.

Interactive fuzzy multiobjective linear programming

Step 0: Calculate the individual minimum and maximum of each objective function under the given constraints.

Step 1*: Elicit a membership function from the DM for each of the objective functions.

Step 2: Set the initial reference membership levels to 1.

Step 3: For the reference membership levels, solve the corresponding minimax problem to obtain the M-Pareto optimal solution and the membership function value together with the trade-off rate information between the membership functions.

Step 4*: If the DM is satisfied with the current levels of the M-Pareto optimal solution, stop. Then the current M-Pareto optimal solution is the satisficing solution for the DM. Otherwise, ask the DM to update the current reference membership levels by considering the current values of the membership functions together with the trade-off rates between the membership functions and return to Step 3.

It should be stressed to the DM that any improvement of one membership function can be achieved only at the expense of at least one of the other membership functions.

Further details concerning the algorithm, extensions, and applications can be found in Sakawa [78].

2.7 Genetic algorithms

2.7.1 Outline of genetic algorithms

Genetic algorithms [43], initiated by Holland, his colleagues and his students at the University of Michigan in the 1970s, as stochastic search techniques based on the mechanism of natural selection and natural genetics, have received a great deal of attention regarding their potential as optimization techniques for solving discrete optimization problems or other hard optimization problems. Although genetic algorithms were not well known at the beginning, after the publication of Goldberg's book [28], genetic algorithms have recently attracted considerable attention in a number of fields as a methodology for optimization, adaptation and learning [1, 2, 6, 15, 16, 23, 67, 71, 110, 126].

Genetic algorithms start with an initial population of individuals generated at random. Each individual in the population represents a potential solution to the problem under consideration. The individuals evolve through successive iterations, called generations. During each generation, each individual in the population is evaluated using some measure of its fitness. Then the population of the next generation is created through genetic operators. The procedure continues until the termination condition is satisfied. The general framework of genetic algorithms is described as follows [67], where $P(t)$ denotes the population at generation t:

procedure: Genetic Algorithms
begin
  $t := 0$;
  initialize $P(t)$;
  evaluate $P(t)$;
  while (**not** termination condition) **do**
  begin
    $t := t + 1$;
    select $P(t)$ from $P(t-1)$;
    alter $P(t)$;
    evaluate $P(t)$;
  end
end.

To explain how genetic algorithms work for an optimization problem, consider a population which consists of N individuals representing potential solutions to the problem. In genetic algorithms, an n-dimensional vector $\boldsymbol{x}$ of decision variables corresponding to an individual is represented by a string $\boldsymbol{s}$ of length n as follows:

$$\boldsymbol{x} : \boldsymbol{s} = s_1 s_2 \cdots s_j \cdots s_n. \tag{2.51}$$

The string $\boldsymbol{s}$ is regarded as a chromosome which consists of n genes. The character s_j is a gene at the jth locus, and the different values of a gene is called alleles. The chromosome $\boldsymbol{s}$ is called the genotype of an individual, while the $\boldsymbol{x}$ corresponding to $\boldsymbol{s}$ is called the phenotype. Usually, it is assumed to establish a one-to-one correspondence between genotypes and phenotypes. However, depending on the situation, m-to-one and one-to-m correspondences are also useful. In either case, the mapping from phenotypes to genotypes is called a coding, and the mapping from genotypes to phenotypes is called a decoding. The length of a chromosome is fixed at a certain value n in many cases, but a chromosome of variable length is more convenient in some cases.

The fitness is the link between genetic algorithms and the problem to be solved. In maximization problems, the fitness of a string $\boldsymbol{s}$ is usually kept the same as the objective function value $f(\boldsymbol{x})$ of its phenotype $\boldsymbol{x}$. In minimization problems, the fitness of a string $\boldsymbol{s}$ should increase as the objective function value $f(\boldsymbol{x})$ of its phenotype $\boldsymbol{x}$ decreases. Thus, in minimization problems, the string with a smaller objective function value has a higher fitness. Through three main genetic operators together with fitness, the population $P(t)$ at generation t evolves to form the next population $P(t+1)$. After some number of generations, the algorithms converge to the best string $\boldsymbol{s}^*$ hopefully represents the optimal or approximate optimal solution $\boldsymbol{x}^*$ to the optimization problem.

In genetic algorithms, the following three main genetic operators, reproduction, crossover and mutation, are usually used to create the next generation.

Reproduction: According to the fitness values, increase or decrease the number of offsprings for each individual in the population $P(t)$.

Crossover: Select two distinct individuals from the population at random and exchange some portion of the strings between the strings with a probability equal to the crossover rate p_c.

Mutation: Alter one or more genes of a selected individual with a probability equal to the mutation rate p_m.

The fundamental procedures of genetic algorithms can now be summarized as follows:

Fundamental procedures of genetic algorithms

Step 0: (Initialization)
Generate N individuals at random to form the initial population $P(0)$. Set the generation index $t = 0$ and determine the value of the maximal generation T.

Step 1: (Evaluation)
Calculate the fitness value of each individual in the population $P(t)$.

Step 2: (Reproduction)
Apply the reproduction operator to the population $P(t)$.

Step 3: (Crossover)
Apply the crossover operator to the population after reproduction.

Step 4: (Mutation)
Apply the mutation operator to the population after crossover to create the new population $P(t+1)$ of the next generation $t+1$.

Step 5: (Termination test)
If $t = T$, stop. Then an individual with the maximal fitness obtained thus fur is regarded as an approximate optimal solution. Otherwise, set $t = t + 1$ and return to Step 1.

In applying genetic algorithms to solve particular optimization problems, further detailed considerations concerning (1) a genetic representation for potential solutions, (2) a way to create an initial population, (3) an evaluation process in terms of their fitness, (4) genetic operators, (5) constraint-handling techniques, and (6) values for various parameters in genetic algorithms, such as population size, probabilities of applying genetic operators, termination conditions, etc. would be required.

As Goldberg [28] summarized, genetic algorithms differ from conventional optimization and search procedures in the following four ways:

(1) Genetic algorithms work with a coding of the solution set, not the solutions themselves.

(2) Genetic algorithms search from a population of solutions, not a single solution.

(3) Genetic algorithms use fitness information, not derivatives or other auxiliary knowledge.
(4) Genetic algorithms use probabilistic transformation rules, not deterministic ones.

Further details of the theory, methods and applications of genetic algorithms can be found in Goldberg [28], Michalewicz [67], Gen and Cheng [23], and Sakawa and Tanaka [110].

2.7.2 Genetic algorithms with double strings

In this subsection, for convenience in our subsequent discussion, genetic algorithms with double strings for multidimensional 0-1 knapsack problems proposed by Sakawa et al. [82, 96, 97, 98, 106, 107, 109] are briefly explained.

Multidimensional 0-1 knapsack problems. In general, a 0-1 programming problem is formulated as

$$\left.\begin{array}{ll} \text{minimize} & \boldsymbol{cx} \\ \text{subject to} & A\boldsymbol{x} \leq \boldsymbol{b} \\ & x_j \in \{0,1\},\ j=1,\ldots,n \end{array}\right\} \tag{2.52}$$

where $\boldsymbol{c} = (c_1, \ldots, c_n)$, $\boldsymbol{x} = (x_1, \ldots, x_n)^T$, $\boldsymbol{b} = (b_1, \ldots, b_m)^T$, and $A = (a_{ij})$ is an $m \times n$ matrix. For simplicity, it is assumed here that each element of A and $\boldsymbol{b}$ is nonnegative respectively. Then the problem (2.52) can be viewed as a multidimensional 0-1 knapsack problem.

Coding and decoding. For solving 0-1 programming problems through genetic algorithms, an individual is usually represented by a binary 0-1 string [28, 67]. This representation, however, may weaken the ability of genetic algorithms since an individual whose phenotype is feasible is scarcely generated under this representation.

For multidimensional 0-1 knapsack problems, Sakawa et al. [82, 96, 97, 98, 106, 107, 109] proposed a double string representation as shown in Figure 2.12, where $s_{i(j)} \in \{1,0\}$, $i(j) \in \{1,\ldots,n\}$, and $i(j) \neq i(j')$ for $j \neq j'$.

index of variable	$i(1)$	$i(2)$	$\cdots$	$i(n)$
0-1 value	$s_{i(1)}$	$i_{i(2)}$	$\cdots$	$s_{i(n)}$

Figure 2.12. Double string.

In a double string, regarding $i(j)$ and $s_{i(j)}$ as the index of an element in a solution vector and the value of the element respectively, a string **s** can be transformed into a solution $\boldsymbol{x} = (x_1, \ldots, x_n)$ as

$$x_{i(j)} = s_{i(j)},\ j = 1, \ldots, n.$$

Unfortunately, however, since this mapping may generate infeasible solutions, the following decoding algorithm for eliminating infeasible solutions has been proposed [82, 98]. In the algorithm, n, j, $i(j)$, $x_{i(j)}$ and $\boldsymbol{a}_{i(j)}$ denote length of a string, a position in a string, an index of a variable, 0-1 value of a variable with index $i(j)$ decoded from a string and an $i(j)$th column vector of the coefficient matrix A, respectively.

Decoding algorithm for double string

Step 1: Set $j = 1$, $\boldsymbol{\Sigma} = \boldsymbol{0}$.

Step 2: If $s_{i(j)} = 1$, set $j = j+1$ and go to Step 3. Otherwise, i.e., if $s_{i(j)} = 0$, set $j = j + 1$ and go to Step 4.

Step 3: If $\boldsymbol{\Sigma} + \boldsymbol{a}_{s(i)} \leq \boldsymbol{b}$, set $x_{s(i)} = 1$, $\boldsymbol{\Sigma} = \boldsymbol{\Sigma} + \boldsymbol{a}_{i(j)}$ and go to Step 4. Otherwise, set $x_{i(j)} = 0$ and go to Step 4.

Step 4: If $j > n$, stop and regard $\boldsymbol{x} = (x_1, \ldots, x_n)^T$ as phenotype of the individual represented by the double string. Otherwise, return to Step 2.

Fitness and scaling. It seems quite natural to define the fitness function of each individual **s** by

$$f(\mathbf{s}) = \frac{\boldsymbol{c}\boldsymbol{x} - \sum_{j \in J_+} c_j}{\sum_{j \in J_-} c_j - \sum_{j \in J_+} c_j}, \tag{2.53}$$

where **s** denotes an individual represented by a double string and $\boldsymbol{x}$ is the phenotype of **s**. Furthermore, $J_+ = \{j \mid c_j \geq 0, 1 \leq j \leq n\}$ and $J_- = \{j \mid c_j < 0, 1 \leq j \leq n\}$.

Observe that the fitness $f(\mathbf{s})$ becomes as

$$f(\mathbf{s}) = \begin{cases} 0, & \text{if } x_j = 1,\ j \in J_+ \text{ and } x_j = 0,\ j \in J_- \\ 1, & \text{if } x_j = 0,\ j \in J_+ \text{ and } x_j = 1,\ j \in J_- \end{cases}$$

and the fitness $f(\mathbf{s})$ satisfies $0 \leq f(\mathbf{s}) \leq 1$.

In a reproduction operator based on the ratio of fitness of each individual to the total fitness such as an expected value model, it is frequently pointed out that the probability of selection depends on the relative ratio of fitness of each individual. Thus, several scaling mechanisms have been introduced [28, 67]. Here, a linear scaling is adopted. In the linear scaling, fitness f_i of an individual is transformed into f_i' according to

$$f_i' = a \cdot f_i + b, \tag{2.54}$$

where the coefficients a and b are determined so that the mean fitness f_{mean} of the population becomes a fixed point and the maximal fitness f_{max} of the population becomes twice as large as the mean fitness.

Reproduction. Up to now, various reproduction methods have been proposed and considered [28, 67]. Using several multiobjective 0-1 programming test problems, Sakawa et al. [82, 98] investigated the performance of each of the six reproduction operators, i.e., ranking selection, elitist ranking selection, expected value selection, elitist expected value selection, roulette wheel selection and elitist roulette wheel selection, and as a result confirmed that elitist expected value selection is relatively efficient. Based mainly on our experience [82, 98], as a reproduction operator, elitist expected value selection is adopted here. Elitist expected value selection is a combination of elitism and expected value selection as mentioned below.

Elitism: If the fitness of a string in the past populations is larger than that of every string in the current population, preserve this string into the current generation.

Expected value selection: For a population consisting of N strings, the expected value of the number of the ith string $\mathbf{s}_i$ in the next population

$$N_i = \frac{f(\mathbf{s}_i)}{\sum_{i=1}^{N} f(\mathbf{s}_i)} \times N$$

is calculated. Then, the integral part of N_i denotes the deterministic number of the string $\mathbf{s}_i$ preserved in the next population. While, the decimal part of N_i is regarded as probability for one of the string $\mathbf{s}_i$ to survive, i.e., $N - \sum_{i=1}^{N} N_i$ strings are determined on the basis of this probability.

Crossover and mutation. If a single-point or multi-point crossover operator is applied to individuals represented by double strings, an index $i(j)$ in an offspring may take the same number that an index $i(j')$ $(j \neq j')$ takes.

Recall that the same violation occurs in solving traveling salesman problems or scheduling problems through genetic algorithms. One possible approach to circumvent such violation, a crossover method called partially matched crossover (PMX) is useful. The PMX was first proposed by Goldberg and Lingle [29] for tackling a blind traveling salesman problem. It enables us to generate desirable offsprings without changing the double string structure unlike the ordinal representation [30]. However, in order to process each element $s_{i(j)}$ in the double string structure efficiently, it is necessary to modify some points of the procedures. The PMX for double strings can be described as follows:

Partially Matched Crossover (PMX) for double string

Step 1: For two individuals $\mathbf{s}_1$ and $\mathbf{s}_2$ represented by double strings, choose two crossover points.

Step 2: According to the PMX, reorder upper strings of $\mathbf{s}_1$ and $\mathbf{s}_2$ together with the corresponding lower strings which yields $\mathbf{s}_1'$ and $\mathbf{s}_2'$.

Step 3: Exchange lower substrings between two crossover points of $\mathbf{s}_1'$ and $\mathbf{s}_2'$ for obtaining the resulting offsprings $\mathbf{s}_1''$ and $\mathbf{s}_2''$ after the revised PMX for double strings.

It is well recognized that a mutation operator plays a role of local random search in genetic algorithms. Here, for the lower string of a double string, mutation of bit-reverse type is adopted. We may introduce another genetic operator, an inversion, together with PMX operator. The inversion proceeds as follows:

Step 1: For an individual $\mathbf{s}$, choose two inversion points at random.

Step 2: Invert both upper and lower substrings between two inversion points.

Termination conditions. When applying genetic algorithms to the multidimensional 0-1 knapsack problem (2.52), an approximate solution of desirable precision must be obtained in a proper time. For this reason, two parameters $I_{\min}$ and $I_{\max}$, which respectively denote the number of generations to be searched at least and at most, are introduced. Then the following termination conditions are imposed.

Step 1: Set the iteration (generation) index $t = 0$ and the parameter of the termination condition $\varepsilon > 0$.

Step 2: Carry out a series of procedures for search through genetic algorithms (reproduction, crossover and mutation).

Step 3: Calculate the mean fitness f_{mean} and the maximal fitness f_{max} of the population.

Step 4: If $t > I_{\text{min}}$ and $(f_{\text{max}} - f_{\text{mean}})/f_{\text{max}} < \varepsilon$, stop.

Step 5: If $t > I_{\text{max}}$, stop. Otherwise, set $t = t + 1$ and return to Step 2.

The readers interested in further details of genetic algorithms with double strings are referred to Sakawa et al. [82, 96, 97, 98, 106, 107, 109, 110].

It is significant to point out here that the genetic algorithms with double strings for 0-1 programming problems of the knapsack type have already been extended by the authors to deal with more general 0-1 programming problems involving both positive and negative coefficients in the constraints. A successful generalization along this line can be found in Sakawa et al. [95, 99, 100], and the interested readers might refer to them for details.

3. The Dantzig-Wolfe Decomposition Method

In this chapter, for convenience in our subsequent discussions in Chapters 4, 5 and 6, the Dantzig-Wolfe decomposition method for large scale linear programming problems with block angular structures is explained in detail. The basic procedure and some of its variants are also introduced.

3.1 Linear programming problems with block angular structures

Consider a large scale linear programming problem which has the following special structure:

$$\left.\begin{array}{lrcl} \text{minimize} & z = \sum_{i=1}^{p} \boldsymbol{c}_i \boldsymbol{x}_i & & \\ \text{subject to} & \sum_{i=1}^{p} A_i \boldsymbol{x}_i & = & \boldsymbol{b}_0 \\ & B_i \boldsymbol{x}_i & = & \boldsymbol{b}_i, \quad i = 1, \ldots, p \\ & \boldsymbol{x}_i & \geq & \boldsymbol{0}, \quad i = 1, \ldots, p \end{array}\right\} \tag{3.1}$$

or equivalently

$$\left.\begin{array}{lcccccccc} \text{minimize} & \boldsymbol{c}_1 \boldsymbol{x}_1 & + & \boldsymbol{c}_2 \boldsymbol{x}_2 & + & \cdots & + & \boldsymbol{c}_p \boldsymbol{x}_p & \\ \text{subject to} & A_1 \boldsymbol{x}_1 & + & A_2 \boldsymbol{x}_2 & + & \cdots & + & A_p \boldsymbol{x}_p & \leq \boldsymbol{b}_0 \\ & B_1 \boldsymbol{x}_1 & & & & & & & \leq \boldsymbol{b}_1 \\ & & & B_2 \boldsymbol{x}_2 & & & & & \leq \boldsymbol{b}_2 \\ & & & & & \ddots & & & \vdots \\ & & & & & & & B_p \boldsymbol{x}_p & \leq \boldsymbol{b}_p \\ & \boldsymbol{x}_j \geq \boldsymbol{0}, \ j = 1, \ldots, p & & & & & & & \end{array}\right\} \tag{3.2}$$

where $\boldsymbol{c}_i$, $i = 1, \ldots, p$, are n_i dimensional row vectors, $\boldsymbol{x}_i$, $i = 1, \ldots, p$, are n_i dimensional decision variable column vectors, A_i, $i = 1, \ldots, p$, are $m_0 \times n_i$

coefficient matrices, B_i, $i = 1,\ldots,p$, are $m_i \times n_i$, and $\boldsymbol{b}_i$, $i = 0,1,\ldots,p$, are m_i dimensional column vectors. The number of variables in the problem (3.1) is $\sum_{i=1}^{p} n_i$ and the number of constraints is $\sum_{i=0}^{p} m_i$. The constraints $B_i\boldsymbol{x}_i = \boldsymbol{b}_i$ including only a certain $\boldsymbol{x}_i$ are called the block constraints, while the constraints $\sum_{i=1}^{p} A_i\boldsymbol{x}_i = \boldsymbol{b}_0$ involving all of $\boldsymbol{x}_i$ are called the coupling constraints.

The structure of this problem is shown in Figure 3.1.

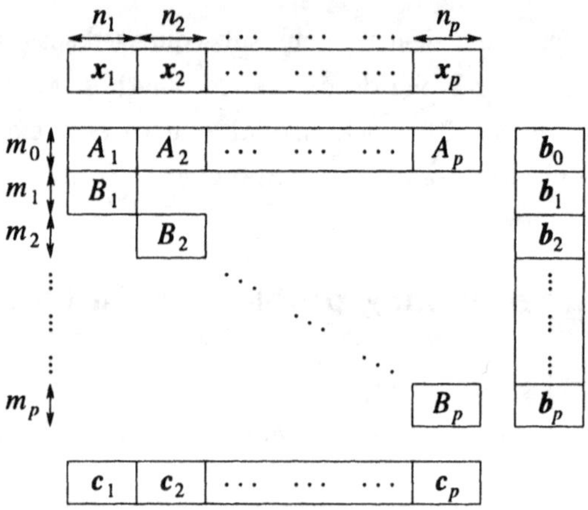

Figure 3.1. Block angular structure.

Problems with such a structure are said to have block angular structures, which are often seen in actual, large scale linear programming problems. For example, consider a production planning problem in a manufacturing company having a number of divisions, where each division has its own limited amounts of internal resources and the divisions are coupled by limited amounts of shared resources. Such a production planning problem can be formulated as a linear programming problem with the block angular structure expressed by (3.1).

Naturally, it is theoretically possible to solve the linear programming problem (3.1) by the revised simplex method using the $\sum_{i=0}^{p} m_i \times \sum_{i=0}^{p} m_i$ basis matrix. However, considering the special structure of the constraints, reduction of both the processing time and memory requirements can be expected by dividing the problem (3.1) into p independent subproblems expressed by

$$\left.\begin{array}{lrcl} \text{minimize} & \boldsymbol{c}_i \boldsymbol{x}_i & & \\ \text{subject to} & B_i \boldsymbol{x}_i & = & \boldsymbol{b}_i \\ & \boldsymbol{x}_i & \geq & \boldsymbol{0}. \end{array}\right\} \tag{3.3}$$

It should be noted that the combination of the p optimal solutions corresponding to the p subproblems is not always a feasible solution of (3.1) because of the existence of the coupling constraints $\sum_{i=1}^{p} A_i \boldsymbol{x}_i = \boldsymbol{b}_0$.

In the early 1960s, Dantzig and Wolfe [13, 14] proposed the Dantzig-Wolfe decomposition method which, after breaking down the original problem (3.1) into one master problem and p independent subproblems, obtains the optimal solution by solving the master problem and these subproblems alternatingly.

To explain the Dantzig-Wolfe decomposition method, first assume that the convex polytope defined by

$$K_i = \{\boldsymbol{x}_i \mid B_i \boldsymbol{x}_i = \boldsymbol{b}_i,\ \boldsymbol{x}_i \geq \boldsymbol{0}\} \tag{3.4}$$

is nonempty and bounded, i.e., a convex polyhedron. If the convex polyhedron K_i has l_i extreme points denoted by $\boldsymbol{v}_i^\ell$, $\ell = 1, \ldots, l_i$, the well-known theorem of the convex polyhedra says that any point $\boldsymbol{x}_i \in K_i$, which satisfies the ith block constraint and the nonnegativity condition, can be expressed as the following convex combination of $\boldsymbol{v}_i^\ell$:

$$\left.\begin{array}{rcl} \boldsymbol{x}_i & = & \displaystyle\sum_{\ell=1}^{l_i} \lambda_i^\ell \boldsymbol{v}_i^\ell \\ \displaystyle\sum_{\ell=1}^{l_i} \lambda_i^\ell & = & 1, \quad i = 1, \ldots, p \\ \lambda_i^\ell & \geq & 0, \quad i = 1, \ldots, p,\ \ell = 1, \ldots, l_i. \end{array}\right\} \tag{3.5}$$

Now, for convenience, assume that all of $\boldsymbol{v}_i^\ell$ have been obtained and define $\boldsymbol{p}_i^\ell$ and f_i^ℓ as

$$\left.\begin{array}{rcll} \boldsymbol{p}_i^\ell & = & A_i \boldsymbol{v}_i^\ell, & \ell = 1, \ldots, l_i \\ f_i^\ell & = & \boldsymbol{c}_i \boldsymbol{v}_i^\ell, & i = 1, \ldots, p. \end{array}\right\} \tag{3.6}$$

Then the original problem (3.1) can be formally rewritten as the following linear programming problem (3.7) with the decision variables λ_i^ℓ:

$$\left.\begin{array}{ll} \text{minimize} & z = \displaystyle\sum_{i=1}^{p}\sum_{\ell=1}^{l_i} f_i^\ell \lambda_i^\ell \\ \text{subject to} & \displaystyle\sum_{i=1}^{p}\sum_{\ell=1}^{l_i} \boldsymbol{p}_i^\ell \lambda_i^\ell = \boldsymbol{b}_0 \\ & \displaystyle\sum_{\ell=1}^{l_i} \lambda_i^\ell = 1,\ i = 1,\dots,p \\ & \lambda_i^\ell \geq 0,\ i = 1,\dots,p,\ \ell = 1,\dots,l_i. \end{array}\right\} \quad (3.7)$$

This problem is called the master problem which is equivalent to the original, i.e., when optimal solutions $\lambda_i^\ell = \lambda_i^{\ell^*}$, $i = 1, ..., p$, to the master problem (3.7) are obtained, an optimal solution $\boldsymbol{x}_i = \boldsymbol{x}_i^*$ to the original problem (3.1) can be obtained by substituting $\lambda_i^\ell = \lambda_i^{\ell^*}$ into the equation (3.5).

The structure of the master problem (3.7) is shown in Figure 3.2.

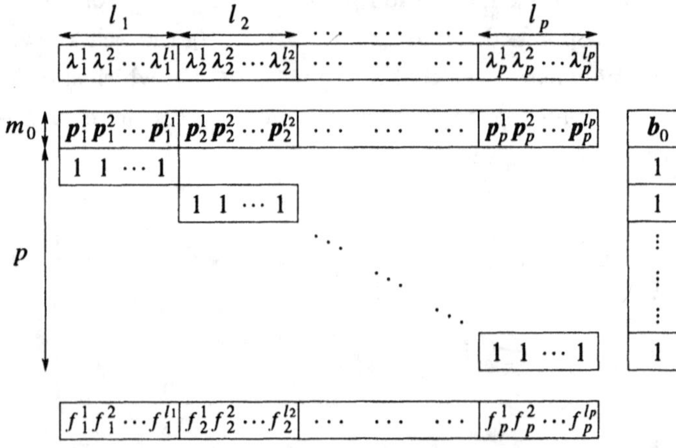

Figure 3.2. Structure of master problem.

If the original problem (3.1) is transformed to the master problem (3.7), the number of constraints decreases from $\sum_{i=0}^{p} m_i$ to m_0+p but the number of variables drastically increases from $\sum_{i=1}^{p} n_i$ to $\sum_{i=1}^{p} l_i$. Moreover, all extreme points of the convex polyhedron K_i should be computed before transforming into the master problem (3.7). This computation requires a formidable effort and solving the master problem is more complicated than solving the original problem (3.1) even if all extreme points are obtained. However, in the Dantzig-Wolfe decomposition method, without the computation of all extreme points, the master problem (3.7) can be solved if simplex multipliers with respect to a basic feasible solution of the master problem (3.7) are given.

3.2 Development of the decomposition algorithm

Suppose an initial feasible basic solution of the master problem (3.7) is given. The corresponding basis matrix B to the feasible basic solution $\boldsymbol{\lambda}_B$ of the master problem includes at least one extreme point in each of the convex polyhedra K_i, $i = 1, \ldots, p$.

Then

$$\boldsymbol{\lambda}_B = B^{-1} \begin{pmatrix} \boldsymbol{b}_0 \\ 1 \\ 1 \\ \vdots \\ 1 \end{pmatrix} \geq \mathbf{0}, \tag{3.8}$$

and let the simplex multipliers for the matrix B be defined by

$$\boldsymbol{\pi} = (\boldsymbol{\pi}_0, \bar{\boldsymbol{\pi}}), \tag{3.9}$$

where $\boldsymbol{\pi}_0$ and $\bar{\boldsymbol{\pi}}$ denote the m_0 dimensional row vector corresponding to the coupling constraints in the master problem and the p dimensional row vector corresponding to the block constraints, respectively.

In order to know if the current feasible basic solution $\boldsymbol{\lambda}_B$ is optimal, we have only to see whether all of the following relative cost coefficients $\bar{f}_i^\ell$ for the nonbasic variables are nonnegative or not:

$$\bar{f}_i^\ell = f_i^\ell - \boldsymbol{\pi} \begin{pmatrix} \boldsymbol{p}_i^\ell \\ \boldsymbol{e}_i \end{pmatrix}, \tag{3.10}$$

where $\boldsymbol{e}_i$ is the p dimensional unit column vector whose ith element is 1 and all the others are 0. The column vectors in Figure 3.2 can be written as $((\boldsymbol{p}_i^\ell)^T, (\boldsymbol{e}_i)^T)^T$.

By defining

$$\bar{\boldsymbol{\pi}} = (\bar{\pi}_1, \bar{\pi}_2, \ldots, \bar{\pi}_p), \tag{3.11}$$

the relative cost coefficients $\bar{f}_i^\ell$ can be expressed as

$$\bar{f}_i^\ell = f_i^\ell - \boldsymbol{\pi}_0 \boldsymbol{p}_i^\ell - \bar{\pi}_i. \tag{3.12}$$

Moreover, from (3.6), it is rewritten as

$$\bar{f}_i^\ell = (\boldsymbol{c}_i - \boldsymbol{\pi}_0 A_i)\boldsymbol{v}_i^\ell - \bar{\pi}_i. \tag{3.13}$$

If the minimal value of the relative cost coefficients $\bar{f}_i^\ell$ for all the nonbasic variables λ_i^ℓ is nonnegative, i.e., the conditions

$$\min_i \min_\ell \bar{f}_i^\ell \geq 0, \quad i = 1, \ldots, p, \ell = 1, \ldots, l_i. \tag{3.14}$$

hold, the current solution is optimal.

Note that $\min_i \min_\ell \bar{f}_i^\ell$ is expressed as

$$\begin{aligned} \min_i \min_\ell \bar{f}_i^\ell &= \min_i \min_\ell \left\{ (\boldsymbol{c}_i - \boldsymbol{\pi}_0 A_i)\, \boldsymbol{v}_i^\ell - \bar{\pi}_i \right\} \\ &= \min_{1 \leq i \leq p} \left\{ \left(\min_{1 \leq \ell \leq l_i} (\boldsymbol{c}_i - \boldsymbol{\pi}_0 A_i)\, \boldsymbol{v}_i^\ell \right) - \bar{\pi}_i \right\}. \end{aligned} \tag{3.15}$$

Then by defining the indices $\ell^*(i) \in \{1, 2, \ldots, l_i\}$, $i = 1, \ldots, p$, by

$$(\boldsymbol{c}_i - \boldsymbol{\pi}_0 A_i)\, \boldsymbol{v}_i^{\ell^*(i)} - \bar{\pi}_i = \min_{1 \leq \ell \leq l_i} (\boldsymbol{c}_i - \boldsymbol{\pi}_0 A_i)\, \boldsymbol{v}_i^\ell - \bar{\pi}_i, \tag{3.16}$$

we can compute

$$\min_i \min_\ell \bar{f}_i^\ell = (\boldsymbol{c}_{i^*} - \boldsymbol{\pi}_0 A_{i^*})\, \boldsymbol{v}_{i^*}^{\ell^*(i^*)} - \bar{\pi}_{i^*} = \min_{1 \leq i \leq p} \left\{ (\boldsymbol{c}_i - \boldsymbol{\pi}_0 A_i)\, \boldsymbol{v}_i^{\ell^*(i)} - \bar{\pi}_i \right\}. \tag{3.17}$$

Since an optimal solution of a linear programming problem, whose constraint sėt is bounded, always occurs at an extreme point of the constraint set, finding the $\boldsymbol{v}_i^{\ell^*(i)}$ in (3.16) is equivalent to solving the modified subproblem defined by

$$\left.\begin{array}{rl} \text{minimize} & (\boldsymbol{c}_i - \boldsymbol{\pi}_0 A_i)\, \boldsymbol{x}_i \\ \text{subject to} & B_i \boldsymbol{x}_i = \boldsymbol{b}_i \\ & \boldsymbol{x}_i \geq \boldsymbol{0}. \end{array}\right\} \tag{3.18}$$

Using the minimal objective function values $z_i^0(\boldsymbol{\pi}_0)$ obtained by solving the modified subproblems (3.18), we have

$$\min_i \min_\ell \bar{f}_i^\ell = \min_{1 \leq i \leq p} \left(z_i^0(\boldsymbol{\pi}_0) - \bar{\pi}_i \right). \tag{3.19}$$

Then, the current solution is optimal if the inequality

$$\min_i \min_\ell \bar{f}_i^\ell = \min_{1 \leq i \leq p} \left(z_i^0(\boldsymbol{\pi}_0) - \bar{\pi}_i \right) \geq 0 \tag{3.20}$$

holds. If not, the column to enter the basis is the column corresponding to the i^* given by

$$z_{i^*}^0(\boldsymbol{\pi}_0) - \bar{\pi}_{i^*} = \min_{1 \le i \le p} \left(z_i^0(\boldsymbol{\pi}_0) - \bar{\pi}_i \right). \tag{3.21}$$

Since $(\boldsymbol{c}_{i^*} - \boldsymbol{\pi}_0 A_{i^*})\, \boldsymbol{v}_{i^*}^{\ell^*(i^*)} - \bar{\pi}_{i^*}$ is the minimal value of $(\boldsymbol{c}_i - \boldsymbol{\pi}_0 A_i)\, \boldsymbol{v}_i^{\ell^*(i)} - \bar{\pi}_i$, $i = 1, \ldots, p$, the vector

$$\begin{pmatrix} \boldsymbol{p}_{i^*}^{\ell^*(i^*)} \\ \boldsymbol{e}_{i^*} \end{pmatrix} = \begin{pmatrix} A_{i^*} \boldsymbol{x}_{i^*}(\boldsymbol{\pi}_0) \\ \boldsymbol{e}_{i^*} \end{pmatrix} \tag{3.22}$$

is the column to enter the basis, i.e., the variable $\lambda_{i^*}^{\ell^*(i^*)}$ becomes a basic variable, where $\boldsymbol{x}_{i^*}(\boldsymbol{\pi}_0)$ $(= \boldsymbol{v}_{i^*}^{\ell^*(i^*)})$ is the solution of the i^*th subproblem.

Based on the preceding discussions, the algorithm of the Dantzig-Wolfe decomposition method is summarized as follows:

The Dantzig-Wolfe decomposition method

Step 1: Assume that an initial feasible basic solution for the master problem (3.7) is available, with a basis matrix B and simplex multipliers $(\boldsymbol{\pi}_0, \bar{\pi}_1, \ldots, \bar{\pi}_p)$.

Step 2: Solve the following p independent subproblems, modified by the simplex multipliers $\boldsymbol{\pi}_0$:

$$\left.\begin{array}{ll} \text{minimize} & (\boldsymbol{c}_i - \boldsymbol{\pi}_0 A_i)\, \boldsymbol{x}_i \\ \text{subject to} & B_i \boldsymbol{x}_i = \boldsymbol{b}_i \\ & \boldsymbol{x}_i \ge \boldsymbol{0} \end{array}\right\}, \quad i = 1, 2, \ldots, p. \tag{3.23}$$

Then, from their solutions $\boldsymbol{x}_i(\boldsymbol{\pi}_0)$ and the corresponding objective function values $z_i^0(\boldsymbol{\pi}_0)$, calculate

$$z_i^0(\boldsymbol{\pi}_0) - \bar{\pi}_i. \tag{3.24}$$

In this step, the sensitivity analysis for linear programming can be used to solve each of the subproblems (3.23), where the coefficients in its objective function are changed.

Step 3: In the master problem, find

$$\min_i \min_\ell \bar{f}_i^\ell = \min_{1 \le i \le p} \left(z_i^0(\boldsymbol{\pi}_0) - \bar{\pi}_i \right), \tag{3.25}$$

where $z_i^0(\boldsymbol{\pi}_0) - \bar{\pi}_i$, $i = 1, \ldots, p$, are obtained by solving the following p independent subproblems. Then, if

$$\min_{1 \le i \le p} \left(z_i^0(\boldsymbol{\pi}_0) - \bar{\pi}_i \right) \ge 0 \tag{3.26}$$

holds, the current solution given by

$$\boldsymbol{x}_i^0 = \sum_{\ell:\ \text{basic}} \lambda_i^\ell \boldsymbol{v}_i^\ell, \tag{3.27}$$

is optimal, where $\boldsymbol{v}_i^\ell$ are the extreme points corresponding to λ_i^ℓ in K_i.

Step 4: If the minimum of $(z_i^0(\boldsymbol{\pi}_0)-\bar{\pi}_i)$ is $(z_{i^*}^0(\boldsymbol{\pi}_0)-\bar{\pi}_{i^*})$, and $(z_{i^*}^0(\boldsymbol{\pi}_0)-\bar{\pi}_{i^*}) < 0$ holds for the solution $\boldsymbol{x}_{i^*}(\boldsymbol{\pi}_0)$ $(= \boldsymbol{v}_{i^*}^{\ell^*(i^*)})$ of the i^*th subproblem, form the column

$$\boldsymbol{p} = \begin{pmatrix} A_{i^*}\boldsymbol{x}_{i^*}(\boldsymbol{\pi}_0) \\ \boldsymbol{e}_{i^*} \end{pmatrix} \tag{3.28}$$

and multiply it by B^{-1} to obtain

$$\bar{\boldsymbol{p}} = B^{-1} \begin{pmatrix} A_{i^*}\boldsymbol{x}_{i^*}(\boldsymbol{\pi}_0) \\ \boldsymbol{e}_{i^*} \end{pmatrix}. \tag{3.29}$$

Then, according to the usual simplex algorithm, perform a pivot operation so that $\lambda_{i^*}^{\ell^*(i^*)}$ enters to the basis. As a result, the inverse matrix B^{-1} of the new basis matrix B and the new simplex multipliers $(\boldsymbol{\pi}_0, \bar{\pi}_1, \ldots, \bar{\pi}_p)$ are obtained. Return to Step 2.

3.3 Initial feasible basic solution

In order to find an initial feasible basic solution, phase one of the simplex method is also effective. Then, the Dantzig-Wolfe decomposition principle can be also applied.

3.4 Unbounded subproblem

Consider the case in which the convex polytope for the ith block constraint

$$K_i = \{\boldsymbol{x}_i \mid B_i\boldsymbol{x}_i = \boldsymbol{b}_i, \boldsymbol{x}_i \geq \boldsymbol{0}\} \tag{3.30}$$

is unbounded. In this case, Using l_i extreme points $\{\boldsymbol{v}_i^1, \ldots, \boldsymbol{v}_i^{l_i}\}$ of K_i and h_i points $\{\boldsymbol{w}_i^1, \ldots, \boldsymbol{w}_i^{h_i}\}$ on h_i extreme rays of a convex cone

$$C_i = \{\boldsymbol{x}_i \mid B_i\boldsymbol{x}_i = \boldsymbol{0}, \boldsymbol{x}_i \geq \boldsymbol{0}\}, \tag{3.31}$$

any point $\boldsymbol{x}_i \in K_i$ is represented by

$$\left.\begin{array}{l} x_i = \sum_{\ell=1}^{l_i} \lambda_i^\ell v_i^\ell + \sum_{h=1}^{h_i} \mu_i^h w_i^h \\ \sum_{\ell=1}^{l_i} \lambda_i^\ell = 1, \ i = 1, \ldots, p \\ \quad \lambda_i^\ell \geq 0, \ i = 1, \ldots, p, \ \ell = 1, \ldots, l_i \\ \sum_{\ell=h}^{h_i} \mu_i^h = 1, \ i = 1, \ldots, p \\ \quad \mu_i^h \geq 0, \ i = 1, \ldots, p, \ h = 1, \ldots, h_i. \end{array}\right\} \quad (3.32)$$

Here, for the nonbasic variables whose relative cost coefficients are negative in solving the ith subproblem, when each element $\bar{a}_{js}$ of a transformed column vector $\bar{p}_s = (\bar{a}_{1s}, \ldots, \bar{a}_{m_i s})^T$ is nonpositive, i.e., the solution is unbounded, w_i^h is calculated from the column vector $\bar{p}_s$ as follows:

$$w_i^h = \begin{pmatrix} -\bar{p}_s \\ e_s \end{pmatrix}, \quad (3.33)$$

where e_s is the $(n_i - m_i)$ dimensional unit column vector whose sth element is equal to 1 and others are all 0. Now, in addition to (3.6), introducing

$$\left.\begin{array}{lll} q_i^h & = A_i w_i^h, & h = 1, \ldots, h_i \\ g_i^h & = c_i w_i^h, & i = 1, \ldots, p \end{array}\right\}, \quad (3.34)$$

the original problem can be formally written as the following linear programming problem with decision variables λ_i^ℓ and μ_i^h.

$$\left.\begin{array}{ll} \text{minimize} & \sum_{i=1}^{p} \left\{ \sum_{\ell=1}^{l_i} f_i^\ell \lambda_i^\ell + \sum_{h=1}^{h_i} g_i^h \mu_i^h \right\} \\ \text{subject to} & \sum_{i=1}^{p} \left\{ \sum_{\ell=1}^{l_i} p_i^\ell \lambda_i^\ell + \sum_{h=1}^{h_i} q_i^h \mu_i^h \right\} = b_0 \\ & \qquad \sum_{\ell=1}^{l_i} \lambda_i^\ell = 1, \ i = 1, \ldots, p \\ & \qquad \lambda_i^\ell \geq 0, \ i = 1, \ldots, p, \ \ell = 1, \ldots, l_i \\ & \qquad \mu_i^h \geq 0, \ i = 1, \ldots, p, \ h = 1, \ldots, h_i \end{array}\right\} \quad (3.35)$$

Figure 3.3 illustrates the structure of the master problem (3.35) when subproblems are unbounded.

Since the relative cost coefficients of the nonbasic variables μ_i^h:

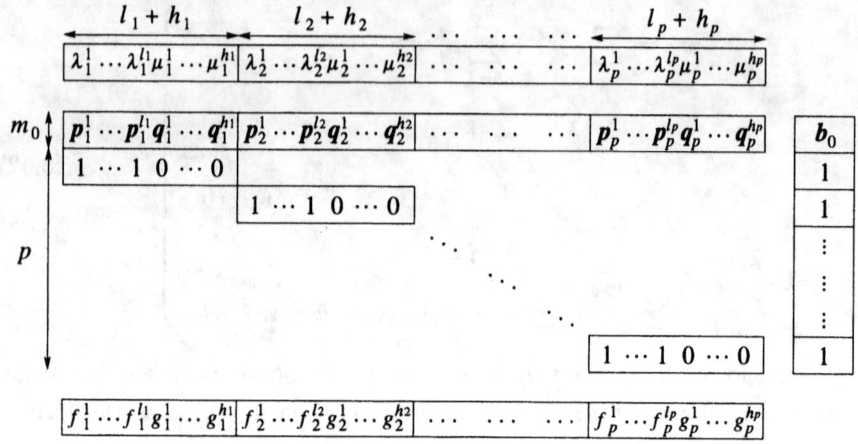

Figure 3.3. Structure of master problem for unbounded subproblems.

$$\begin{aligned} \bar{g}_i^h &= g_i^h - \pi \begin{pmatrix} q_i^h \\ 0 \end{pmatrix} \\ &= g_i^h - \pi_0 q_i^h \\ &= (c_i - \pi_0 A_i)\, w_i^h \end{aligned} \tag{3.36}$$

are always negative, the columns in the form of $((A_i w_i^h)^T, 0^T)^T$, corresponding to μ_i^h, are candidates to enter the basis.

3.5 Restricted master problem

As discussed thus far, the Dantzig-Wolfe decomposition method solves p independent linear programming problems in the subproblems, but in the master problem only a single pivot operation is performed. By defining the restricted master problem, a more symmetric formulation is possible in which both levels solve linear programming problems.

Observe that, in Step 4 of the Dantzig-Wolfe decomposition algorithm, new simplex multipliers $(\pi_0, \bar{\pi}_1, \ldots, \bar{\pi}_p)$ are obtained by constructing the column $((A_s x_s(\pi_0))^T, e_s^T)^T$ for a solution $x_s(\pi_0)$ of the sth subproblem and making B^{-1} enter the basis, while a column $((A_i x_i(\pi_0))^T, e_i^T)^T$ can be constructed from the solution $x_i(\pi_0)$ of another subproblem. It is likely that relative cost coefficients are negative for the previous simplex multipliers in most of these columns. If, in any of those columns, the relative cost coefficient calculated using the new simplex multipliers remains negative (this often occurs because the simplex multipliers don't change seriously), these columns can be used again to decrease the objective values.

For these reasons, the following restricted master problem is considered by adding new columns $\boldsymbol{p}_1^*, \boldsymbol{p}_2^*, \ldots, \boldsymbol{p}_p^*$, corresponding to the solutions of the subproblems, to the column set in the current basis $\{\boldsymbol{p}_i^\ell\}$:

$$\left.\begin{array}{lrcll} \text{minimize} & \displaystyle\sum_{i=1}^{p} \sum_{\ell:\text{ basic}} f_i^\ell \lambda_i^\ell + \sum_{i=1}^{p} f_i^* \lambda_i^* & & & \\ \text{subject to} & \displaystyle\sum_{i=1}^{p} \sum_{\ell:\text{ basic}} \boldsymbol{p}_i^\ell \lambda_i^\ell + \sum_{i=1}^{p} \boldsymbol{p}_i^* \lambda_i^* & = & \boldsymbol{b}_0 & \\ & \displaystyle\sum_{\ell:\text{ basic}} \lambda_i^\ell + \lambda_i^* & = & 1, & i = 1, \ldots, p \\ & \lambda_i^\ell & \geq & 0, & i = 1, \ldots, p \\ & \lambda_i^* & \geq & 0, & i = 1, \ldots, p \end{array}\right\} \qquad (3.37)$$

Noting that the number of variables = the number of constraints + p, a larger decrease of the objective value can be expected by solving the restricted master problem, where p basis transformations are done simultaneously.

Now, if the optimality conditions (3.20) are not satisfied by the solutions of the subproblems for the simplex multipliers obtained in the restricted master problem, by removing p variables except for the optimal basic variables of the previous restricted master problem and adding p variables corresponding to the p optimal solutions of the p subproblems, a new restricted master problem with $m_0 + 2p$ variables is constructed. Using the simplex multipliers for the optimal solution of the new restricted master problem, the p subproblems are solved again. These procedures are iterated until the optimality conditions (3.20) are satisfied.

For the computational issues related to actually implementing the Dantzig-Wolfe decomposition method, the readers are referred to Ho and Loute [38, 39], and Ho and Sundarraj [40, 41]. Especially, the monograph of Ho and Sundarraj [42] has the complete documentation of DECOMP: a robust implementation of the Dantzig-Wolfe decomposition method in FORTRAN, and the code can serve as a very convenient starting point for further investigation.

Observe that the paper by Todd [132] gives a straightforward description of a variant of Karmarkar's projective algorithm [45] that avoids projective transformations and bears a strong resemblance to the Dantzig-Wolfe decomposition method.

A recently published book by Martin [66] which covers the major theoretical and practical advances in the wide range of large scale linear and integer linear programming is very useful for interested readers.

4. Large Scale Fuzzy Linear Programming

This chapter treats large scale linear programming problems with block angular structures for which the Dantzig-Wolfe decomposition method has been successfully applied. Considering the vague or fuzzy nature of human judgments, both the fuzzy goal and fuzzy constrains of the decision maker are introduced. Having determined the corresponding membership functions, following the convex fuzzy decision for combining them, under suitable conditions, it is shown that the formulated problem can be reduced to a number of independent linear subproblems and the satisficing solution for the decision maker is directly obtained just only by solving the subproblems. Moreover, even if the appropriate conditions are not satisfied, it is shown that the Dantzig-Wolfe decomposition method is applicable.

4.1 Linear programming problems with block angular structures

Similar to the previous chapter, consider a large scale linear programming problem of the following block angular form:

$$\left.\begin{array}{lrlllll} \text{minimize} & \boldsymbol{c}\boldsymbol{x} = & \boldsymbol{c}\boldsymbol{x}_1 & + \cdots + & \boldsymbol{c}\boldsymbol{x}_p & & \\ \text{subject to} & A\boldsymbol{x} = & A_1\boldsymbol{x}_1 & + \cdots + & A_p\boldsymbol{x}_p & \leq & \boldsymbol{b}_0 \\ & & B_1\boldsymbol{x}_1 & & & \leq & \boldsymbol{b}_1 \\ & & & \ddots & & & \vdots \\ & & & & B_p\boldsymbol{x}_p & \leq & \boldsymbol{b}_p \\ & & \multicolumn{5}{l}{\boldsymbol{x}_j \geq \boldsymbol{0},\ j = 1, \ldots, p} \end{array}\right\} \tag{4.1}$$

where $\boldsymbol{c}_i$, $i = 1, \ldots, p$, are n_i dimensional cost factor row vectors, $\boldsymbol{x}_i$, $i = 1, \ldots, p$, are n_i dimensional vectors of decision variables, $A\boldsymbol{x} = A_1\boldsymbol{x}_1 + \cdots + A_p\boldsymbol{x}_p \leq \boldsymbol{b}_0$ are m_0 dimensional coupling constraints, A_i, $i = 1, \ldots, p$, are $m_0 \times n_i$ coefficient matrices. $B_i\boldsymbol{x}_i \leq \boldsymbol{b}_i$, $i = 1, \ldots, p$, are m_i dimensional block constraints with respect to $\boldsymbol{x}_i$ and B_i, $i = 1, \ldots, p$, are $m_i \times n_i$ coefficient matrices.

We can interpret this problem in terms of an overall system with p subsystems. The overall system is composed of p subsystems and have an objective function $\boldsymbol{cx}$, which is the sum of separable objective functions $\boldsymbol{c}_i\boldsymbol{x}$, $i = 1, \cdots, p$, for each subsystem i. There are coupling constraints and constraints for each subsystem.

Note that if the coupling constraints do not exist, the overall problem can be decomposed into the following p independent linear programming subproblems with coefficient matrix B_i, $i = 1, \ldots, p$:

$$\left.\begin{array}{rl} \text{minimize} & \boldsymbol{c}_i\boldsymbol{x}_i \\ \text{subject to} & B_i\boldsymbol{x}_i \leq \boldsymbol{b}_i, \ \boldsymbol{x}_i \geq \boldsymbol{0}. \end{array}\right\} \tag{4.2}$$

Theoretically, it is possible to solve the problem by directly applying the revised simplex method for $\sum_{i=0}^{p} m_i \times \sum_{i=0}^{p} m_i$ basis matrix.

However the larger the overall system becomes, the more difficult it becomes to solve due to the high dimensionality of the system. To circumvent this difficulty, the Dantzig-Wolfe decomposition method [13, 14, 35, 36, 37, 38, 39, 40, 41, 42, 58] has been proposed by utilizing its special structure.

4.2 Fuzzy goal and fuzzy constraints

In 1976, Zimmermann [141] first introduced the fuzzy set theory into conventional linear programming problems. He considered linear programming problems with a fuzzy goal and fuzzy constraints. Following the fuzzy decision proposed by Bellman and Zadeh [3] together with linear membership functions, he proved that there exists an equivalent linear programming problem. Since then, fuzzy linear programming has been developed in a number of directions with many successful applications [17, 44, 56, 73, 74, 78, 128, 134, 144, 145].

By considering the vague nature of human judgments for large scale linear programming problems with block angular structures, it is quite natural to assume that the decision maker (DM) may have a fuzzy goal for the objective function and fuzzy constrains for the coupling constraints. These fuzzy goal and fuzzy constraints can be quantified by eliciting the corresponding membership functions through the interaction with the DM.

To elicit a linear membership function $\mu_0(\boldsymbol{cx})$ from the DM for a fuzzy goal, the DM is asked to assess a minimum value of unacceptable levels for

$\boldsymbol{cx}$, denoted by z_0^0, and a maximum value of totally desirable levels for $\boldsymbol{cx}$, denoted by z_0^1. Then the linear membership function $\mu_0(\boldsymbol{cx})$ for the fuzzy goal of the DM is defined by

$$\mu_0(\boldsymbol{cx}) = \begin{cases} 1, & \boldsymbol{cx} \leq z_0^1, \\ \dfrac{\boldsymbol{cx} - z_0^0}{z_0^1 - z_0^0}, & z_0^1 \leq \boldsymbol{cx} \leq z_0^0, \\ 0, & \boldsymbol{cx} \geq z_0^0. \end{cases} \tag{4.3}$$

Similarly, for each of the coupling constraints, by assessing a minimum value of unacceptable levels, denoted by z_j^0 and a maximum value of totally desirable levels, denoted by z_j^1, the following linear membership functions $\mu_j((\boldsymbol{Ax})_j)$, $j = 1, \ldots, m_0$, can be determined:

$$\mu_j((\boldsymbol{Ax})_j) = \begin{cases} 1, & (\boldsymbol{Ax})_j \leq z_j^1, \\ \dfrac{(\boldsymbol{Ax})_j - z_j^0}{z_j^1 - z_j^0}, & z_j^1 \leq (\boldsymbol{Ax})_j \leq z_j^0, \\ 0, & (\boldsymbol{Ax})_j \geq z_j^0, \end{cases} \tag{4.4}$$

where $(\boldsymbol{Ax})_j$ denotes the jth component of the column vector $\boldsymbol{Ax}$.

These membership functions are depicted in Figures 4.1 and 4.2.

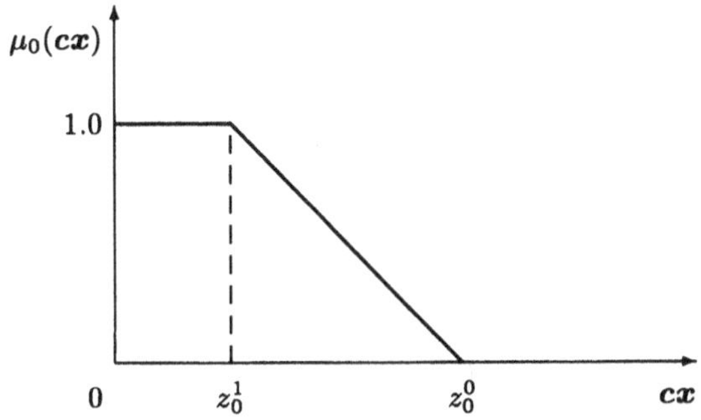

Figure 4.1. Linear membership function for fuzzy goal.

In the followings, for notational convenience, if we set

$$\alpha_j = \frac{1}{z_j^1 - z_j^0}, \ j = 0, 1, \ldots, m_0 \tag{4.5}$$

and

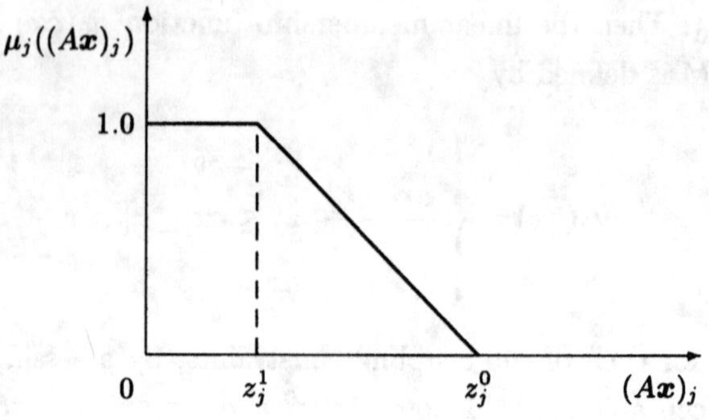

Figure 4.2. Linear membership function for fuzzy constraints.

$$\beta_j = \frac{z_j^0}{z_j^1 - z_j^0}, \; j = 0, 1, \ldots, m_0, \tag{4.6}$$

these membership functions are simply expressed as

$$\mu_0(\boldsymbol{cx}) = \begin{cases} 1, & \boldsymbol{cx} \leq z_0^1, \\ \alpha_0 \boldsymbol{cx} + \beta_0, & z_0^1 \leq \boldsymbol{cx} \leq z_0^0, \\ 0, & \boldsymbol{cx} \geq z_0^0, \end{cases} \tag{4.7}$$

and

$$\mu_j((A\boldsymbol{x})_j) = \begin{cases} 1, & (A\boldsymbol{x})_j \leq z_j^1, \\ \alpha_j (A\boldsymbol{x})_j + \beta_j, & z_j^1 \leq (A\boldsymbol{x})_j \leq z_j^0, \\ 0, & (A\boldsymbol{x})_j \geq z_j^0, \end{cases} \tag{4.8}$$

for $j = 1, \ldots, m_0$.

4.3 Fuzzy linear programming

Now we are ready to incorporate the fuzzy goal and fuzzy constraints of the decision maker (DM) into large scale linear programming problems with block angular structures.

Realizing that both the fuzzy goal and fuzzy constraints are desired to be satisfied simultaneously, Bellman and Zadeh [3] defined the fuzzy decision resulting from the fuzzy goal and fuzzy constraints as the intersection of them.

However, depending on the situations, other aggregation patterns for the fuzzy goal and the fuzzy constraints may be worth considering. When a fuzzy goal and fuzzy constraints have unequal importance, Bellman and Zadeh

[3] also indicated to use the convex fuzzy decision. As an example of an alternative definition of a fuzzy decision, the product fuzzy decision has also been proposed [142].

In 1976, Zimmermann [141] first considered linear programming problems with a fuzzy goal and fuzzy constraints. Following the fuzzy decision or the minimum operator proposed by Bellman and Zadeh together with linear membership functions, he proved that there exists an equivalent linear programming problem.

In 1978, Sommer and Pollatschek [131] proposed to adopt the add-operator for aggregating the DM's fuzzy goal and fuzzy constraints instead of the minimum operator. Their add-operator can be viewed as a special case of the convex fuzzy decision. Probably the most crucial problem is the identification of an appropriate aggregate operator which well represents the DM's fuzzy preferences [20, 62, 146].

In this section, as a first attempt for incorporating the fuzzy goal and fuzzy constraints into large scale linear programming problems with block angular structures, we adopt the convex fuzzy decision for combining them. Then it can be shown that, under some appropriate conditions, the formulated problem can be reduced to a number of independent linear subproblems and the overall satisficing solution for the DM is directly obtained just only by solving the subproblems [79, 103].

By adopting the convex fuzzy decision, large scale linear programming problems incorporating the fuzzy goal and fuzzy constraints can be formulated as follows:

$$\left.\begin{array}{ll}\text{maximize} & w_0\mu_0(\boldsymbol{cx}) + \sum_{j=1}^{m_0} w_j\mu_j((A\boldsymbol{x})_j) \\ \text{subject to} & B_i\boldsymbol{x}_i \leq \boldsymbol{b}_i,\ \boldsymbol{x}_i \geq \boldsymbol{0},\ i = 1,\ldots,p, \\ & \boldsymbol{cx} \leq z_0^0, \\ & (A\boldsymbol{x})_j \leq z_j^0,\ j = 1,\ldots,m_0\end{array}\right\} \tag{4.9}$$

where $w_j \geq 0$, $j = 0,\ldots,m_0$; $\sum_{j=0}^{m_0} w_j = 1$.

In this formulation observe that the last two constraints are added since each z_j^0, $j = 0,\ldots,m_0$, denotes the minimum value of the unacceptable level assessed by the DM. As a result, although the coupling constraints $A\boldsymbol{x} \leq \boldsymbol{b}$ are incorporated into the objective function through fuzzy constraints, newly added constraints still remain as coupling constraints.

To overcome such difficulties, by eliminating the third and fourth constraints from the problem (4.9), assume the membership functions μ_j, $j =$

$0, \ldots, m_0$, are not line segments within $[z_j^1, z_j^0]$ but straight lines beyond $[z_j^1, z_j^0]$, i.e., $-\alpha_0 \boldsymbol{cx} - \beta_0$ and $-\alpha_j(A\boldsymbol{x})_j - \beta_j$, $j = 1, \ldots, m_0$. Then the problem (4.9) reduces to

$$\left.\begin{array}{ll} \text{maximize} & w_0\left(\alpha_0\left(\sum_{i=1}^{p} \boldsymbol{c}_i\boldsymbol{x}_i\right) + \beta_0\right) + \sum_{j=1}^{m_0} w_j\left(\alpha_j\left(\sum_{i=1}^{p}(A_i\boldsymbol{x}_i)_j\right) + \beta_j\right) \\ \text{subject to} & B_i\boldsymbol{x}_i \leq \boldsymbol{b}_i,\ \boldsymbol{x}_i \geq \boldsymbol{0},\ i = 1, \ldots, p. \end{array}\right\} \tag{4.10}$$

Obviously, in the problem (4.10), there are no coupling constraints and it can be solved by decomposing into the following p independent linear programming subproblems:

$$\left.\begin{array}{ll} \text{maximize} & w_0\alpha_0\boldsymbol{c}_i\boldsymbol{x}_i + \sum_{j=1}^{m_0} w_j\alpha_j(A_i\boldsymbol{x}_i)_j \\ \text{subject to} & B_i\boldsymbol{x}_i \leq \boldsymbol{b}_i,\ \boldsymbol{x}_i \geq \boldsymbol{0}, \end{array}\right\} \tag{4.11}$$

where, for simplicity, constants $w_j\beta_j$, $j = 0, \ldots, m_0$, are omitted.

The relationships between the optimal solutions of the problems (4.10) and (4.9) can be given in the following theorem.

Theorem 4.3.1. *Let $\hat{\boldsymbol{x}}_i$, $i = 1, 2, \ldots, p$, be optimal solutions to the p linear subproblems* (4.11). *Then if $\hat{\boldsymbol{x}}_i$, $i = 1, 2, \ldots, p$, satisfy*

$$z_0^1 \leq \sum_{i=1}^{p} \boldsymbol{c}_i\hat{\boldsymbol{x}}_i \leq z_0^0, \tag{4.12}$$

and

$$z_j^1 \leq (\sum_{i=1}^{p} A_i\hat{\boldsymbol{x}}_i)_j \leq z_j^0,\ j = 1, 2, \ldots, m_0, \tag{4.13}$$

then $\hat{\boldsymbol{x}} = (\hat{\boldsymbol{x}}_1^T, \hat{\boldsymbol{x}}_2^T, \ldots, \hat{\boldsymbol{x}}_p^T)^T$ is an optimal solution to the problem (4.9).

Proof. Since $\hat{\boldsymbol{x}}$ is an optimal solution to the problem (4.10) satisfying (4.12) and (4.13), $\hat{\boldsymbol{x}}$ is an optimal solution to the following problem:

$$\left.\begin{array}{ll} \text{maximize} & w_0\left(\alpha_0\left(\sum_{i=1}^{p} \boldsymbol{c}_i\boldsymbol{x}_i\right) + \beta_0\right) + \sum_{j=1}^{m_0} w_j\left(\alpha_j\left(\sum_{i=1}^{p}(A_i\boldsymbol{x}_i)_j\right) + \beta_j\right) \\ \text{subject to} & B_i\boldsymbol{x}_i \leq \boldsymbol{b}_i,\ \boldsymbol{x}_i \geq \boldsymbol{0},\ i = 1, \ldots, p, \\ & \sum_{i=1}^{p} \boldsymbol{c}_i\boldsymbol{x}_i \leq z_0^0, \\ & \sum_{i=1}^{p}(A_i\boldsymbol{x}_i)_j \leq z_j^0, j = 1, \ldots, m_0. \end{array}\right\} \tag{4.14}$$

Let $\tilde{\boldsymbol{x}} = (\tilde{\boldsymbol{x}}_1^T, \tilde{\boldsymbol{x}}_2^T, \dots, \tilde{\boldsymbol{x}}_p^T)^T$ be an optimal solution to the problem (4.9), then from the last two constraints of the problem (4.9), it holds that

$$\alpha_0 \left(\sum_{i=1}^{p} \boldsymbol{c}_i \tilde{\boldsymbol{x}}_i \right) + \beta_0 \geq \mu_0 \left(\sum_{i=1}^{p} \boldsymbol{c}_i \tilde{\boldsymbol{x}}_i \right),$$

$$\alpha_j \left(\sum_{i=1}^{p} (A_i \tilde{\boldsymbol{x}}_i)_j \right) + \beta_j \geq \mu_j \left(\sum_{i=1}^{p} (A_i \tilde{\boldsymbol{x}}_i)_j \right), \ j = 1, \dots, m_0.$$

In view of (4.12) and (4.12), it follows that

$$\begin{aligned} & w_0 \mu_0 \left(\sum_{i=1}^{p} \boldsymbol{c}_i \hat{\boldsymbol{x}}_i \right) + \sum_{j=1}^{m_0} w_j \mu_j \left(\sum_{i=1}^{p} (A_i \hat{\boldsymbol{x}}_i)_j \right) \\ & = w_0 \left(\alpha_0 \left(\sum_{i=1}^{p} \boldsymbol{c}_i \hat{\boldsymbol{x}}_i \right) + \beta_0 \right) + \sum_{j=1}^{m_0} \left(w_j \alpha_j \left(\sum_{i=1}^{p} (A_i \hat{\boldsymbol{x}}_i)_j \right) + \beta_j \right) \\ & \geq w_0 \left(\alpha_0 \left(\sum_{i=1}^{p} \boldsymbol{c}_i \tilde{\boldsymbol{x}}_i \right) + \beta_0 \right) + \sum_{j=1}^{m_0} \left(w_j \alpha_j \left(\sum_{i=1}^{p} (A_i \tilde{\boldsymbol{x}}_i)_j \right) + \beta_j \right) \\ & \geq w_0 \mu_0 \left(\sum_{i=1}^{p} \boldsymbol{c}_i \tilde{\boldsymbol{x}}_i \right) + \sum_{j=1}^{m_0} w_j \mu_j \left(\sum_{i=1}^{p} (A_i \tilde{\boldsymbol{x}}_i)_j \right), \end{aligned}$$

which implies $\hat{\boldsymbol{x}}$ is an optimal solution to the problem (4.9)

Theorem 4.3.1 means that if optimal solutions to the p subproblems (4.11) satisfy the conditions (4.12) and (4.13), then the satisficing solution for the DM for large scale linear programming problems incorporating the fuzzy goal and fuzzy constraints can be derived by just solving the p independent linear programming subproblems.

The remaining problem is to consider the case when optimal solutions to the p subproblems (4.11) do not satisfy the conditions (4.12) and (4.13).

For that purpose, observe that if any feasible solution $\boldsymbol{x}$ of the problem (4.9) satisfies

$$\boldsymbol{c}\boldsymbol{x} \geq z_0^1, \tag{4.15}$$

and

$$(A\boldsymbol{x})_j \geq z_j^1, \ j = 1, \dots, m_0, \tag{4.16}$$

then the problem (4.9) becomes equivalently to the large scale linear programming problem with the block angular structure (4.14) and hence the Dantzig-Wolfe decomposition method is applicable. One possible way to guarantee these conditions (4.15) and (4.15) is to determine z_0^1 and z_j^1, $j = 1, \dots, m_0$ as is given in the following theorem.

Theorem 4.3.2. *Let* z_0^1 *be an optimal value to the linear programming problem*

$$\left.\begin{array}{rl} \text{minimize} & \sum_{i=1}^{p} \boldsymbol{c}_i \boldsymbol{x}_i \\ \text{subject to} & B_i \boldsymbol{x}_i \leq \boldsymbol{b}_i,\ \boldsymbol{x}_i \geq \mathbf{0},\ i = 1, \ldots, p \end{array}\right\} \tag{4.17}$$

and z_j^1, $j = 1, \ldots, m_0$, *be an optimal value to the linear programming problem*

$$\left.\begin{array}{rl} \text{minimize} & \sum_{i=1}^{p} (A_i \boldsymbol{x}_i)_j \\ \text{subject to} & B_i \boldsymbol{x}_i \leq \boldsymbol{b}_i,\ \boldsymbol{x}_i \geq \mathbf{0},\ i = 1, \ldots, p. \end{array}\right\} \tag{4.18}$$

Then any feasible solution $\boldsymbol{x}$ *of the problem* (4.9) *satisfies conditions* (4.15) *and* (4.16).

Proof. Since (4.17) and (4.18) are part of the constraints of the problem (4.9), it is evident that any feasible solution $\boldsymbol{x}$ to the problem (4.9) satisfies (4.15) and (4.16).

From Theorem 4.3.2, even if (4.12) and (4.13) are not satisfied, by setting z_0^1 and z_j^1, $j = 1, \ldots, m_0$, as the optimal values of the problem (4.17) and (4.18), the Dantzig-Wolfe decomposition method is applicable for solving the problem (4.9).

It should be noted here that the problems (4.17) and (4.18) can be decomposed respectively into the following p independent linear subproblems:

$$\left.\begin{array}{rl} \text{minimize} & \boldsymbol{c}_i \boldsymbol{x}_i \\ \text{subject to} & B_i \boldsymbol{x}_i \leq \boldsymbol{b}_i,\ \boldsymbol{x}_i \geq \mathbf{0} \end{array}\right\} \tag{4.19}$$

and

$$\left.\begin{array}{rl} \text{minimize} & \sum_{i=1}^{p} (A_i \boldsymbol{x}_i)_j \\ \text{subject to} & B_i \boldsymbol{x}_i \leq \boldsymbol{b}_i,\ \boldsymbol{x}_i \geq \mathbf{0}. \end{array}\right\} \tag{4.20}$$

Hence the determination of z_0^1 and z_j^1, $j = 1, \ldots, m_0$, as is shown in Theorem 4.3.2 can be easily done.

With the convex fuzzy decision, it is now appropriate to consider some reasonable way to determine the weighting coefficients w_j, $j = 0, \ldots, m_0$. To do so, it seems to be quite natural to assume that the DM has satisfactory levels z_j^{f}, $j = 0, \ldots, m_0$, for the objective function and coupling constraints since in general totally desirable levels z_j^1, $j = 0, \ldots, m_0$, as was introduced in Theorem 4.3.2 are not simultaneously attainable. From the definition of z_j^{f}

it is evident to assume $z_j^1 < z_j^{\mathrm{f}} < z_j^0$, $j = 0, \ldots, m_0$. Then one possible way to determine the weighting coefficients w_j, $j = 0, \ldots, m_0$, is to select them so that the relations $w_i \mu_i(z_i^{\mathrm{f}}) = w_j \mu_j(z_j^{\mathrm{f}})$, $i, j \in \{0, \ldots, m_0\}$ hold, i.e.,

$$w_j = \frac{\dfrac{1}{\mu_j(z_j^{\mathrm{f}})}}{\displaystyle\sum_{i=0}^{s} \frac{1}{\mu_i(z_i^{\mathrm{f}})}}, \quad j = 0, \ldots, m_0. \tag{4.21}$$

Following the preceding discussions, we can now construct the algorithm in order to derive the satisficing solution for the DM to the large scale linear programming problem with the block angular structure incorporating the fuzzy goal and fuzzy constraints.

Fuzzy linear programming

Step 1: Calculate z_0^1 and z_j^1, $j = 1, \ldots, m_0$, by solving the independent linear subproblems (4.19) and (4.20).

Step 2: With information of z_0^1 and z_j^1, $j = 1, \ldots, m_0$, elicit linear membership functions from the DM for the objective function and coupling constraints by assessing the minimum values of the unacceptable levels z_0^0 and z_j^0, $j = 1, \ldots, m_0$.

Step 3: Determine the weighting coefficients w_j, $j = 0, \ldots, m_0$, according to the relations (4.21) by assessing the satisfactory levels z_j^{f}, $j = 0, \ldots, m_0$.

Step 4: Solve p linear subproblems (4.11) for obtaining the optimal solutions $\hat{\boldsymbol{x}} = (\hat{\boldsymbol{x}}_1^T, \hat{\boldsymbol{x}}_2^T, \ldots, \hat{\boldsymbol{x}}_p^T)^T$ to the problem (4.10).

Step 5: Check whether $\hat{\boldsymbol{x}}$ satisfies (4.12) and (4.13) or not. If (4.12) and (4.13) are satisfied, stop. Then the current optimal solution $\hat{\boldsymbol{x}}$ is the satisficing solution for the DM. Otherwise, proceed to the next step.

Step 6: Solve the large scale linear programming problem (4.14) by applying the Dantzig-Wolfe decomposition method for obtaining the optimal solution $\tilde{\boldsymbol{x}}$. Then the current optimal solution $\tilde{\boldsymbol{x}}$ is the satisficing solution for the DM.

It is shown that the satisficing solution for the DM to large scale linear programming problems with the block angular structure incorporating the fuzzy goal and fuzzy constraints can be obtained by solving a number of linear subproblems. Especially, if conditions (4.12) and (4.13) hold, the overall satisficing solution is directly derived just only by solving p independent linear subproblems without using the Dantzig-Wolfe decomposition method. Even if (4.12) and (4.13) do not hold, the problem can be solved by applying the Dantzig-Wolfe decomposition method. Observe that the larger the spreads $z_j^0 - z_j^1$ of the fuzzy goal and fuzzy constraints are, the more often (4.12) and (4.13) hold. Thus, the proposed method will be still more powerful when the DM has very vague goal and constraints.

4.4 Numerical example

To illustrate the proposed method, consider the following simple numerical example [58]:

$$\left.\begin{array}{rrcrcrcrl} \text{minimize} & -x_1 & - & x_2 & - & 2y_1 & - & y_2 & \\ \text{subject to} & x_1 & + & 2x_2 & + & 2y_1 & + & y_2 & \leq 40 \\ & x_1 & + & 3x_2 & & & & & \leq 30 \\ & 2x_1 & + & x_2 & & & & & \leq 20 \\ & & & & & y_1 & & & \leq 10 \\ & & & & & & & y_2 & \leq 10 \\ & & & & & y_1 & + & y_2 & \leq 15 \\ & \multicolumn{3}{r}{x_1, x_2 \geq 0,} & & \multicolumn{4}{l}{y_1, y_2 \geq 0} \end{array}\right\}$$

Observe that this problem has one coupling constraint and two independent blocks, and a step-by-step procedure for solving the problem using the Dantzig-Wolfe decomposition method can be found in [58].

First solving

$$\left.\begin{array}{rrcrcrcrl} \text{minimize} & -x_1 & - & x_2 & - & 2y_1 & - & y_2 & \\ \text{subject to} & x_1 & + & 3x_2 & & & & & \leq 30 \\ & 2x_1 & + & x_2 & & & & & \leq 20 \\ & & & & & y_1 & & & \leq 10 \\ & & & & & & & y_2 & \leq 10 \\ & & & & & y_1 & + & y_2 & \leq 15 \\ & \multicolumn{3}{r}{x_1, x_2 \geq 0,} & & \multicolumn{4}{l}{y_1, y_2 \geq 0} \end{array}\right\}$$

by decomposing into 2 independent linear subproblems, yields $z_0^1 = -44$.

Similarly, solving

$$\left.\begin{array}{lrcrcrcrl}
\text{minimize} & x_1 & + & x_2 & + & 2y_1 & + & y_2 & \\
\text{subject to} & x_1 & + & 3x_2 & & & & & \leq 30 \\
& 2x_1 & + & x_2 & & & & & \leq 20 \\
& & & & & y_1 & & & \leq 10 \\
& & & & & & & y_2 & \leq 10 \\
& & & & & y_1 & + & y_2 & \leq 15 \\
& x_1, x_2 \geq 0, & & & & y_1, y_2 \geq 0 & & &
\end{array}\right\}$$

z_1^1 becomes 0.

By considering these values, assume that a hypothetical decision maker (DM) assessed the values of z_0^0 and z_1^0 to be -34 and 50, respectively. Then the corresponding membership functions $\mu_0(\boldsymbol{cz})$ and $\mu_1(\boldsymbol{az})$ become as follows:

$$\mu_0(\boldsymbol{cz}) = \begin{cases} 1, & \boldsymbol{cz} \leq -44, \\ -\dfrac{\boldsymbol{cz} - 34}{10}, & -34 \leq \boldsymbol{cz} \leq -44, \\ 0, & \boldsymbol{cz} \geq -34, \end{cases}$$

$$\mu_1(\boldsymbol{az}) = \begin{cases} 1, & \boldsymbol{az} \leq 0, \\ -\dfrac{\boldsymbol{az} - 50}{50}, & 0 \leq \boldsymbol{az} \leq 50, \\ 0, & \boldsymbol{az} \geq 50, \end{cases}$$

where $\boldsymbol{c} = (\boldsymbol{c}_1, \boldsymbol{c}_2)$, $\boldsymbol{a} = (\boldsymbol{a}_1, \boldsymbol{a}_2)$ and $\boldsymbol{z} = (\boldsymbol{x}^T, \boldsymbol{y}^T)^T$.

These membership functions are respectively depicted in Figures 4.3 and 4.4.

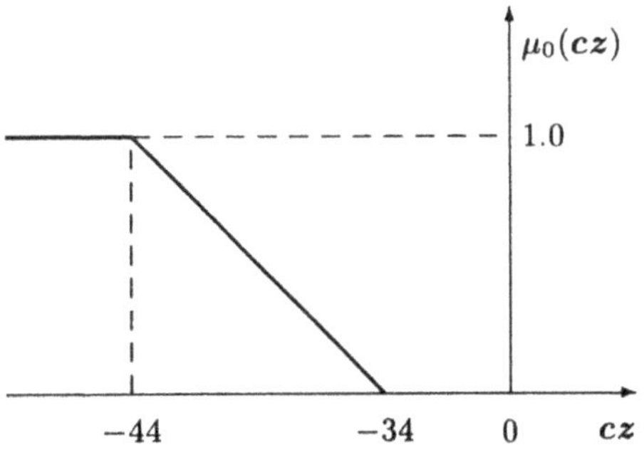

Figure 4.3. Linear membership function for fuzzy goal.

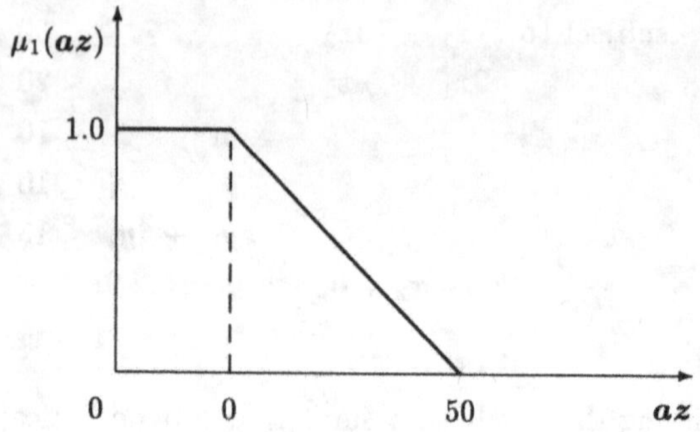

Figure 4.4. Linear membership function for fuzzy constraint.

Also assume that a hypothetical DM assessed the values of z_0^{f} and z_1^{f} to be -37 and 40, respectively. Then calculating the weights w_0 and w_1 according to (4.21), yields $w_0 = 2/5$ and $w_1 = 3/5$.

Consequently, solving two linear programming subproblems

$$\left.\begin{array}{ll} \text{maximize} & -\dfrac{1}{25}\boldsymbol{c}_1\boldsymbol{x} - \dfrac{3}{250}\boldsymbol{a}_1\boldsymbol{x} \\ \text{subject to} & x_1 + 3x_2 \leq 30, \\ & 2x_1 + x_2 \leq 20, \\ & x_1 \geq 0,\ x_2 \geq 0, \end{array}\right\}$$

and

$$\left.\begin{array}{ll} \text{maximize} & -\dfrac{1}{25}\boldsymbol{c}_2\boldsymbol{y} - \dfrac{3}{250}\boldsymbol{a}_2\boldsymbol{y} \\ \text{subject to} & y_1 \leq 10, \\ & y_2 \leq 10, \\ & y_1 + y_2 \leq 15, \\ & y_1 \geq 0,\ y_2 \geq 0, \end{array}\right\}$$

yields the corresponding optimal solutions $\boldsymbol{x}^* = (6,8)^T$ and $\boldsymbol{y}^* = (10,5)^T$, for which conditions $-44 \leq \boldsymbol{cz} \leq -34$ and $0 \leq \boldsymbol{az} \leq 50$ are satisfied. Therefore, from Theorem 4.3.1, the satisficing solution for the DM becomes $\boldsymbol{x}^* = (6,8)$ and $\boldsymbol{y}^* = (10,5)$. Compared with an optimal solution of the original problem, it should be noted here that incorporation of the fuzzy goal and fuzzy constraint into the original problem improves the objective function value by 2.334 at the expense of the left-hand side value of the coupling constraint by 1.

5. Large Scale Fuzzy Multiobjective Linear Programming

As a multiobjective generalization of the previous chapter, this chapter treats large scale multiobjective linear programming problems with block angular structures. The fuzzy goals of the decision maker are quantified by eliciting the corresponding linear membership functions. Following the fuzzy decision of Bellman and Zadeh for combining them, fuzzy multiobjective linear programming, using the Dantzig-Wolfe decomposition method, is presented for obtaining the satisficing solution for the decision maker. Interactive fuzzy multiobjective linear programming for deriving the satisficing solution for the decision maker by updating the reference membership levels is also presented by utilizing the Dantzig-Wolfe decomposition method.

5.1 Multiobjective linear programming problems with block angular structures

As a multiobjective generalization of the previous chapter, consider a large scale multiobjective linear programming problem of the following block angular form:

$$\left.\begin{array}{rrcllll}
\text{minimize} & \boldsymbol{c}_1\boldsymbol{x} = & \boldsymbol{c}_{11}\boldsymbol{x}_1 & +\cdots+ & \boldsymbol{c}_{1p}\boldsymbol{x}_p & & \\
 & & & \vdots & & & \\
\text{minimize} & \boldsymbol{c}_k\boldsymbol{x} = & \boldsymbol{c}_{k1}\boldsymbol{x}_1 & +\cdots+ & \boldsymbol{c}_{kp}\boldsymbol{x}_p & & \\
\text{subject to} & A\boldsymbol{x} = & A_1\boldsymbol{x}_1 & +\cdots+ & A_p\boldsymbol{x}_p & \leq & \boldsymbol{b}_0 \\
 & & B_1\boldsymbol{x}_1 & & & \leq & \boldsymbol{b}_1 \\
 & & & \ddots & & & \vdots \\
 & & & & B_p\boldsymbol{x}_p & \leq & \boldsymbol{b}_p \\
 & & & \multicolumn{3}{c}{\boldsymbol{x}_j \geq \boldsymbol{0},\ j = 1,\ldots,p} &
\end{array}\right\} \tag{5.1}$$

where $\boldsymbol{c}_{ij}$, $i = 1,\ldots,k$; $j = 1,\ldots,p$, are n_i dimensional cost factor row vectors, $\boldsymbol{x}_j$, $j = 1,\ldots,p$, are n_i dimensional vectors of decision variables,

$A\boldsymbol{x} = A_1\boldsymbol{x}_1 + \cdots + A_p\boldsymbol{x}_p \leq \boldsymbol{b}_0$ are m_0 dimensional coupling constraints, A_j, $j = 1, \ldots, p$, are $m_0 \times n_i$ coefficient matrices. $B_j\boldsymbol{x}_j \leq \boldsymbol{b}_j$, $j = 1, \ldots, p$, are m_i dimensional block constraints with respect to $\boldsymbol{x}_j$ and B_j, $j = 1, \ldots, p$ are $m_i \times n_i$ coefficient matrices. In the followings, for notational convenience, let X denote the feasible region satisfying all of the constraints of the problem (5.1).

We can interpret this problem in terms of an overall system with p subsystems. The overall system is composed of p subsystems and has k objective functions $\boldsymbol{c}_i\boldsymbol{x}$, $i = 1, \ldots, k$, each of which is the sum of separable objective functions $\boldsymbol{c}_{ij}\boldsymbol{x}$, $j = 1, \ldots, p$, for each subsystem j. There are a coupling constraint and constraints for each subsystem.

In general, however, for multiobjective linear programming problems, a complete optimal solution which simultaneously minimizes all of the multiple objective functions does not always exist when the objective functions conflict with each other. Thus, instead of a complete optimal solution, a new solution concept, called Pareto optimality, is introduced in multiobjective linear programming as discussed in Chapter 2.

5.2 Fuzzy goals

In 1978, Zimmermann first introduced the fuzzy set theory into multiobjective linear programming problems [142]. He considered multiobjective linear programming problems with fuzzy goals. Following the fuzzy decision proposed by Bellman and Zadeh [3] together with linear membership functions, he proved that there exists an equivalent linear programming problem. Since then, fuzzy linear programming has been developed in a number of directions with many successful applications [17, 44, 56, 78, 128, 134, 144, 145].

By considering the vague nature of human judgments, for large scale multiobjective linear programming problems with block angular structures, it is quite natural to assume that the decision maker (DM) may have a fuzzy goal for each of the objective functions. These fuzzy goals can be quantified by eliciting the corresponding membership functions through the interaction with the DM.

To elicit a linear membership function $\mu_i(\boldsymbol{c}_i\boldsymbol{x})$ for each i from the DM for each of the fuzzy goals, the DM is asked to assess a minimum value of unacceptable levels for $\boldsymbol{c}_i\boldsymbol{x}$, denoted by z_i^0 and a maximum value of totally desirable levels for $\boldsymbol{c}_i\boldsymbol{x}$, denoted by z_i^1. Then the linear membership functions

$\mu_i(\boldsymbol{c}_i\boldsymbol{x})$, $i = 1,\ldots,k$, for the fuzzy goals of the DM are defined by

$$\mu_i(\boldsymbol{c}_i\boldsymbol{x}) = \begin{cases} 1, & \boldsymbol{c}_i\boldsymbol{x} \leq z_i^1 \\ \dfrac{\boldsymbol{c}_i\boldsymbol{x} - z_i^0}{z_i^1 - z_i^0}, & z_i^1 \leq \boldsymbol{c}_i\boldsymbol{x} \leq z_i^0 \\ 0, \boldsymbol{c}_i\boldsymbol{x} \geq z_i^0. \end{cases} \tag{5.2}$$

These membership functions are depicted in Figure 5.1.

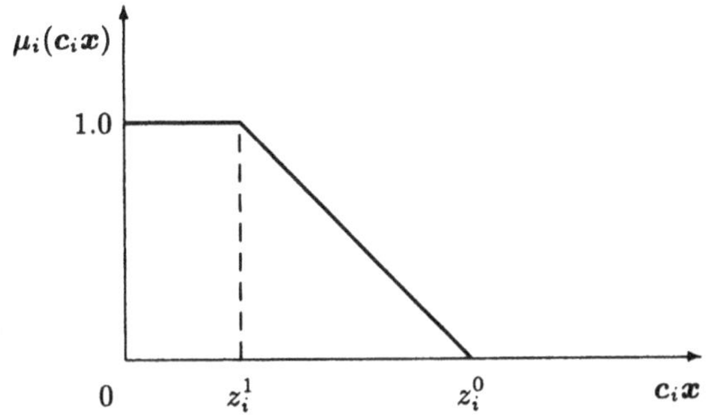

Figure 5.1. Linear membership function for fuzzy goals.

In the followings, for notational convenience, if we set

$$\alpha_i = \frac{1}{z_i^1 - z_i^0}, \quad i = 1,\ldots,k \tag{5.3}$$

and

$$\beta_i = \frac{z_{j,0}^0}{z_i^1 - z_i^0}, \quad i = 1,\ldots,k \tag{5.4}$$

these membership functions are simply expressed as

$$\mu_i(\boldsymbol{c}_i\boldsymbol{x}) = \begin{cases} 1, & \boldsymbol{c}_i\boldsymbol{x} \leq z_0^1 \\ \alpha_i\boldsymbol{c}_i\boldsymbol{x} + \beta_i, & z_i^1 \leq \boldsymbol{c}_i\boldsymbol{x} \leq z_i^0 \\ 0, & \boldsymbol{c}_i\boldsymbol{x} \geq z_i^0. \end{cases} \tag{5.5}$$

As one of the possible ways to help the DM determine z_i^0 and z_i^1, it is convenient to calculate the minimal value $z_i^{\min}$ and the maximal value $z_i^{\max}$ of each objective function under the given constraints. Namely, $z_i^{\min}$ is the optimal value of the linear programming problem

$$\left.\begin{array}{lrlr}
\text{minimize} & \boldsymbol{c}_i\boldsymbol{x} = & \boldsymbol{c}_{j1}\boldsymbol{x}_1 + \cdots + \boldsymbol{c}_{jp}\boldsymbol{x}_p & \\
\text{subject to} & A\boldsymbol{x} = & A_1\boldsymbol{x}_1 + \cdots + A_p\boldsymbol{x}_p & \leq \boldsymbol{b}_0 \\
 & & B_1\boldsymbol{x}_1 & \leq \boldsymbol{b}_1 \\
 & & \ddots & \vdots \\
 & & B_p\boldsymbol{x}_p & \leq \boldsymbol{b}_p \\
 & & \boldsymbol{x}_j \geq \boldsymbol{0},\ j = 1, \ldots, p &
\end{array}\right\} \quad (5.6)$$

and $z_i^{\max}$ is the optimal value of the linear programming problem

$$\left.\begin{array}{lrlr}
\text{maximize} & \boldsymbol{c}_i\boldsymbol{x} = & \boldsymbol{c}_{j1}\boldsymbol{x}_1 + \cdots + \boldsymbol{c}_{jp}\boldsymbol{x}_p & \\
\text{subject to} & A\boldsymbol{x} = & A_1\boldsymbol{x}_1 + \cdots + A_p\boldsymbol{x}_p & \leq \boldsymbol{b}_0 \\
 & & B_1\boldsymbol{x}_1 & \leq \boldsymbol{b}_1 \\
 & & \ddots & \vdots \\
 & & B_p\boldsymbol{x}_p & \leq \boldsymbol{b}_p \\
 & & \boldsymbol{x}_j \geq \boldsymbol{0},\ j = 1, \ldots, p &
\end{array}\right\} \quad (5.7)$$

to both of which the Dantzig-Wolfe decomposition method is applicable.

Then by taking account of the calculated individual minimum and maximum of each objective function, the DM is asked to assess z_i^0 and z_i^1 in the interval $[z_i^{\min}, z_i^{\max}]$, $i = 1, \ldots, k$.

5.3 Fuzzy multiobjective linear programming

Now we are ready to incorporate the fuzzy goals of the decision maker (DM) into large scale multiobjective linear programming problems with block angular structures.

Realizing that all of the fuzzy goals are desired to be satisfied simultaneously, Bellman and Zadeh [3] defined the fuzzy decision resulting from the fuzzy goals as the intersection of them.

In 1978, Zimmermann [142] first considered multiobjective linear programming problems with fuzzy goals. Following the fuzzy decision or minimum operator proposed by Bellman and Zadeh together with linear membership functions, he proved that there exists an equivalent linear programming problem.

In this section, as a first attempt for incorporating the fuzzy goals into large scale multiobjective linear programming problems with block angular structures, we adopt minimum operator for combining them. Then it can be shown that the formulated problem can be reduced to one master problems and a number of independent linear subproblems and the satisficing solution for the DM can be obtained by applying the Dantzig-Wolfe decomposition method.

After eliciting the linear membership functions, if we adopt the minimum operator, large scale multiobjective linear programming problems incorporating the fuzzy goals of the DM can be formulated as

$$\underset{\boldsymbol{x} \in X}{\text{maximize}} \min_{i=1,\ldots,k} \mu_i(\boldsymbol{c}_i \boldsymbol{x}). \tag{5.8}$$

By introducing the auxiliary variable λ, this problem can be equivalently transformed as

$$\left.\begin{array}{ll} \text{maximize} & \lambda \\ \text{subject to} & \mu_i(\boldsymbol{c}_i \boldsymbol{x}) \geq \lambda, \; i = 1, \ldots, k \\ & \boldsymbol{x} \in X. \end{array}\right\} \tag{5.9}$$

Associated with this problem, consider the following problem where all of the membership functions $\mu_i(\boldsymbol{c}_i \boldsymbol{x})$, $i = 1, \ldots, k$, are replaced by $\alpha_i \boldsymbol{c}_i \boldsymbol{x} + \beta_i$.

$$\left.\begin{array}{ll} \text{maximize} & \lambda \\ \text{subject to} & \alpha_i \boldsymbol{c}_i \boldsymbol{x} + \beta_i \geq \lambda, \; i = 1, \ldots, k \\ & \boldsymbol{x} \in X. \end{array}\right\} \tag{5.10}$$

Observe that the membership functions $\mu_i(\boldsymbol{c}_i \boldsymbol{x})$, $i = 1, \ldots, k$, are no more line segments within $[z_i^1, z_i^0]$ but straight lines beyond $[z_i^1, z_i^0]$, i.e., $-\alpha_i \boldsymbol{c}_i \boldsymbol{x} - \beta_0$, $i = 1, \ldots, k$. Obviously this problem preserves linearity as well as the block angular structure and hence the Dantzig-Wolfe decomposition method is applicable.

The relationships between the optimal solutions of the problems (5.9) and (5.10) can be given in the following theorem.

Theorem 5.3.1. *If* $(\boldsymbol{x}^*, \lambda^*)$ *is an optimal solution to the problem* (5.10), *then* $(\boldsymbol{x}^*, \max(0, \min(\lambda^*, 1)))$ *is an optimal solution to the problem* (5.9).

Proof. Using min and max operators, (5.5) can be rewritten as

$$\mu_i(\boldsymbol{c}_i \boldsymbol{x}) = \max(0, \; \min(\alpha_i \boldsymbol{c}_i \boldsymbol{x} + \beta_i, \; 1)).$$

First consider the case of $\lambda^* \geq 1$. Since $(\boldsymbol{x}^*, \lambda^*)$ is an optimal solution to (5.10), it obviously follows that

$$\alpha_i \boldsymbol{c}_i \boldsymbol{x}^* + \beta_i \geq \lambda^*, \; i = 1, \ldots, k,$$

which implies

$$\mu_i(\boldsymbol{c}_i \boldsymbol{x}^*) = \max(0, \min(\alpha_i \boldsymbol{c}_i \boldsymbol{x}^* + \beta_i, 1)) \geq \max(0, \min(\lambda^*, 1)), \; i = 1, \ldots, k.$$

Hence, $(\boldsymbol{x}^*, \max(0, \min(\lambda^*, 1)))$ is a feasible solution to (5.9). In view of $\lambda \leq \mu_i(\boldsymbol{c}_i \boldsymbol{x}) \leq 1$ for all feasible solutions $(\boldsymbol{x}, \lambda)$ to (5.9) together with $\max(0, \min(\lambda^*, 1)) = 1$ in the case of $\lambda^* \geq 1$, it follows that $(\boldsymbol{x}^*, \max(0, \min(\lambda^*, 1)))$ is an optimal solution to (5.9).

In the case of $\lambda^* \leq 0$, it follows that $\max(0, \min(\lambda^*, 1)) = 0$ which implies $(\boldsymbol{x}^*, \max\ (0, \min(\lambda^*, 1)))$ is a feasible solution to (5.9). Assume that there exists a feasible solution $(\boldsymbol{x}^1, \lambda^1)$ such that $\lambda^1 > 0$, then it holds that

$$\mu_i(\boldsymbol{c}_i \boldsymbol{x}^1) \geq \lambda^1, \; i = 1, \ldots, k.$$

In view of

$$\alpha_i \boldsymbol{c}_i \boldsymbol{x} + \beta_i \geq \mu_i(\boldsymbol{c}_i \boldsymbol{x}), \; i = 1, \ldots, k$$

in the range $(0, 1]$ of the membership functions, we have

$$\alpha_i \boldsymbol{c}_i \boldsymbol{x}^1 + \beta_i \geq \lambda^1, \; i = 1, \ldots, k$$

which implies $(\boldsymbol{x}^1, \lambda^1)$ is a feasible solution to (5.9). Hence, we have a feasible solution $(\boldsymbol{x}^1, \lambda^1)$ such that $\lambda^1 > 0 \geq \lambda^*$. This contradicts the optimality of $(\boldsymbol{x}^*, \lambda^*)$ to (5.10), which implies $(\boldsymbol{x}^*, \max(0, \min(\lambda^*, 1)))$ is an optimal solution to (5.9).

In the case of $0 < \lambda^* < 1$, it follows that $\max(0, \min(\lambda^*, 1)) = \lambda^*$ which implies

$$\alpha_i \boldsymbol{c}_i \boldsymbol{x} + \beta_i = \mu_i(\boldsymbol{c}_i \boldsymbol{x}), \; i = 1, \ldots, k$$

in the range $(0, 1)$ of the membership functions. Hence, $(\boldsymbol{x}^*, \max(0, \min(\lambda^*, 1)))$ is a feasible solution to (5.9). Quite similar to the case of $\lambda^* \leq 0$, the supposition of the existence of a feasible solution $(\boldsymbol{x}^1, \lambda^1)$ to the problem (5.9) satisfying $\lambda^1 > \lambda^*$ yields the contradiction.

Theorem 5.3.1 means that the satisficing solution for the DM for large scale multiobjective linear programming problems incorporating the fuzzy goals can be derived by solving the large scale linear programming problem (5.10). Observe that this problem (5.10) preserves the block angular structure and hence the Dantzig-Wolfe decomposition method can be applied.

The following theorem guarantees the weak Pareto optimality of the optimal solution to the problem (5.10) for the original large scale multiobjective linear programming problem with the block angular structure.

Theorem 5.3.2. *If $\boldsymbol{x}^*$ is an optimal solution to the linear programming problem* (5.10) *with the block angular structure, then $\boldsymbol{x}^*$ is a weak Pareto optimal solution to the original large scale multiobjective linear programming problem* (5.1).

Proof. Let $(\boldsymbol{x}^*, \lambda^*)$ be an optimal solution to (5.10). Then it obviously follows that $\boldsymbol{x}^* \in X$. Now if there exists $\boldsymbol{x}^1 \in X$ such that $\boldsymbol{c}_i\boldsymbol{x}^1 < \boldsymbol{c}_i\boldsymbol{x}^*$, $i = 1, \ldots, k$, then from $\alpha_i < 0$, it follows that

$$\alpha_i \boldsymbol{c}_i \boldsymbol{x}^1 + \beta_i > \alpha_i \boldsymbol{c}_i \boldsymbol{x}^* + \beta_i, \; i = 1, \ldots, k.$$

By the monotonicity of the minimum operation, we have

$$\min_{i=1,\ldots,k} (\alpha_i \boldsymbol{c}_i \boldsymbol{x}^1 + \beta_i) > \min_{i=1,\ldots,k} (\alpha_i \boldsymbol{c}_i \boldsymbol{x}^* + \beta_i).$$

If we set

$$\lambda^1 = \min_{i=1,\ldots,k} (\alpha_i \boldsymbol{c}_i \boldsymbol{x}^1 + \beta_i)$$

then $(\boldsymbol{x}^1, \lambda^1)$ is a feasible solution to (5.10) and

$$\lambda^1 > \min_{i=1,\ldots,k} (\alpha_i \boldsymbol{c}_i \boldsymbol{x}^* + \beta_i) \geq \lambda^*$$

which contradicts the optimality of $(\boldsymbol{x}^*, \lambda^*)$ to (5.10). Hence there does not exist $\boldsymbol{x}^1 \in X$ such that $\boldsymbol{c}_i\boldsymbol{x}^1 < \boldsymbol{c}_i\boldsymbol{x}^*$, $i = 1, \ldots, k$, which implies $\boldsymbol{x}^*$ is a weak Pareto optimal solution to the original problem (5.1).

Unfortunately, Theorem 5.3.2 guarantees only the weak Pareto optimality of an optimal solution $\boldsymbol{x}^*$ to (5.10). However, it should be stressed that the satisficing solution for the DM should be selected from the set of Pareto optimal solutions of the original problem (5.1). As a result, it is necessary to perform the Pareto optimality test of $\boldsymbol{x}^*$. In view of the definition of the Pareto optimality, $\boldsymbol{x}^* \in X$ is a Pareto optimal solution if and only if there does not exist another $\boldsymbol{x} \in X$ such that $\boldsymbol{c}_i\boldsymbol{x} \leq \boldsymbol{c}_i\boldsymbol{x}^*$, $i = 1, \ldots, k$, and $\boldsymbol{c}_i\boldsymbol{x} < \boldsymbol{c}_i\boldsymbol{x}^*$ for some j. Therefore, the Pareto optimality test for $\boldsymbol{x}^*$ can be performed by solving the following linear programming problem with the decision variables $\boldsymbol{x}$ and $\boldsymbol{\epsilon}$.

$$\left.\begin{array}{ll} \text{maximize} & \sum_{i=1}^{k} \varepsilon_i \\ \text{subject to} & \mathbf{c}_i \boldsymbol{x} + \varepsilon_i = \mathbf{c}_i \boldsymbol{x}^*, \; i = 1, \ldots, k \\ & \boldsymbol{x} \in X \\ & \boldsymbol{\varepsilon} = (\varepsilon_1, \ldots, \varepsilon_k) \geq \mathbf{0}. \end{array}\right\} \quad (5.11)$$

For the optimal solutions $(\widehat{\boldsymbol{x}}, \widehat{\boldsymbol{\varepsilon}})$ of this linear programming problem, the following two cases should be considered.

(a) $\widehat{\varepsilon}_i = 0$ for all $j \in \{1, \ldots, k\}$.
(b) $\widehat{\varepsilon}_i > 0$ for at least one $j \in \{1, \ldots, k\}$.

Case (a) means that $\boldsymbol{x}^* \in X$ itself is a Pareto optimal solution. Whereas case (b) means that $\boldsymbol{x}^* \in X$ is not a Pareto optimal solution. Therefore, for case (a), $\boldsymbol{x}^*$ becomes the satisficing solution for the DM for the problem (5.1) incorporating the fuzzy goals. However, for case (b), we must seek for the satisficing solution for the DM which is also Pareto optimal.

The following two theorems hold for the optimal solutions $(\widehat{\boldsymbol{x}}, \widehat{\boldsymbol{\varepsilon}})$ to the problem (5.11).

Theorem 5.3.3. *Let* $(\widehat{\boldsymbol{x}}, \widehat{\boldsymbol{\varepsilon}})$ *be an optimal solutions to the problem* (5.11). *Then* $\widehat{\boldsymbol{x}}$ *is a Pareto optimal solution to the problem* (5.1).

Proof. If there exists $\boldsymbol{x}^1 \in X$ such that $\mathbf{c}_i \boldsymbol{x}^1 \leq \mathbf{c}_i \widehat{\boldsymbol{x}}$, $i = 1, \ldots, k$, and $\mathbf{c}_{j^1} \boldsymbol{x}^1 < \mathbf{c}_{j^1} \widehat{\boldsymbol{x}}$ for some $j^1 \in \{1, \ldots, k\}$, then by defining $\boldsymbol{\varepsilon}^1 = (\varepsilon_1^1, \ldots, \varepsilon_k^1)$ as

$$\varepsilon_i^1 = \begin{cases} \widehat{\varepsilon}_i, & j \neq j^1 \\ \widehat{\varepsilon}_i + (\mathbf{c}_{j^1} \widehat{\boldsymbol{x}} - \mathbf{c}_{j^1} \boldsymbol{x}^1), & j = j^1 \end{cases}$$

$(\boldsymbol{x}^1, \boldsymbol{\varepsilon}^1)$ is a feasible solution to the problem (5.11). Hence it follows that

$$\sum_{i=1}^{k} \widehat{\varepsilon}_i < \sum_{i=1}^{k} \varepsilon_i^1$$

which contradicts the fact that $(\widehat{\boldsymbol{x}}, \widehat{\boldsymbol{\varepsilon}})$ is an optimal solution to the problem (5.11).

Theorem 5.3.4. *Let* $(\boldsymbol{x}^*, \lambda^*)$ *and* $(\widehat{\boldsymbol{x}}, \widehat{\boldsymbol{\varepsilon}})$ *be optimal solutions to the problems* (5.10) *and* (5.11) *respectively. Then* $(\widehat{\boldsymbol{x}}, \lambda^*)$ *becomes an optimal solution to the problem* (5.10).

Proof. From the constraints of the problem (5.11), it follows that

$$c_i \hat{x} \leq c_i x^*, \; i = 1, \ldots, k.$$

From $\alpha_i < 0$ and (x^*, λ^*) is a feasible solution to the problem (5.10), it holds that

$$\alpha_i c_i \hat{x} + \beta_i \geq \alpha_i c_i x^* + \beta_i \geq \lambda^*, \; i = 1, \ldots, k,$$

which means $(\hat{x}, \lambda^*)$ is a feasible solution to the problem (5.10). Hence $(\hat{x}, \lambda^*)$ is a feasible solution to the problem (5.10). In view of the optimal value of the problem (5.10) is λ^*, $(\hat{x}, \lambda^*)$ is an optimal solution to the problem (5.10).

From Theorems 5.3.3 and 5.3.4, it can be understood that an optimal solution $\hat{x}$ to the problem (5.11) is a Pareto optimal solution to the original problem (5.1) and $(\hat{x}, \lambda^*)$ is an optimal solution to the problem (5.10). Therefore, in case of (b), instead of x^*, $\hat{x}$, which satisfies the Pareto optimality, becomes the satisficing solution for the DM incorporating the fuzzy goals.

In this way, it is shown that the satisficing solution for the DM to large scale multiobjective linear programming problems with the block angular structure incorporating the fuzzy goals can be efficiently obtained by applying the Dantzig-Wolfe decomposition method.

Following the preceding discussions, we can now construct the algorithm in order to derive the satisficing solution for the DM to the large scale multiobjective linear programming problem with the block angular structure incorporating the fuzzy goals of the DM.

Fuzzy multiobjective linear programming

Step 1: Calculate the individual minimum z_i^0 and maximum z_i^1 of each objective function under the given constraints by applying the Dantzig-Wolfe decomposition method to the problems (5.6) and (5.7).

Step 2: With information of z_i^0 and z_i^1, $i = 1, \ldots, k$, determine linear membership functions for the objective functions by asking the DM to assess the values of z_i^0 and z_i^1 in the interval $[z_i^{\min}, z_i^{\max}]$, $i = 1, \ldots, k$.

Step 3: Solve the linear programming problem (5.10) with the block angular structure by applying the Dantzig-Wolfe decomposition method for obtaining the optimal solution to the problem (5.9).

Step 4: For the obtained optimal solution, solve the Pareto optimality test problem (5.11) by applying the Dantzig-Wolfe decomposition method for obtaining the Pareto optimal solution. Then the current optimal solution is the satisficing solution for the DM.

Example 5.3.1. To illustrate the proposed method, consider the following simple numerical example involving two objective functions, one coupling constraint and three independent blocks:

$$\left.\begin{array}{lrl} \text{minimize} & z_1 = & -w_1 - 2w_2 - x_1 - 2x_2 - 2y_1 - y_2 \\ \text{minimize} & z_2 = & 3w_1 + 8w_2 + 4x_1 + 3x_2 + 6y_1 + 4y_2 \\ \text{subject to} & & w_1 + w_2 + x_1 + 2x_2 + 2y_1 + y_2 \leq 60 \\ & & 2w_1 + 3w_2 \leq 40 \\ & & w_1 + 4w_2 \leq 40 \\ & & 5w_1 + w_2 \leq 61 \\ & & x_1 + 3x_2 \leq 30 \\ & & 2x_1 + x_2 \leq 20 \\ & & y_1 \leq 10 \\ & & y_2 \leq 10 \\ & & y_1 + y_2 \leq 15 \\ & & w_i \geq 0,\ x_i \geq 0,\ y_i \geq 0,\ i = 1, 2. \end{array}\right\} \quad (5.12)$$

To calculate the individual minimum and maximum of each objective function under the given constraints, solving (5.6) and (5.7) corresponding to this problem, we obtain $z_1^{\min} = -62$, $z_2^{\min} = 0$, $z_1^{\max} = 0$, $z_2^{\max} = 214$.

By considering these values, assume that a hypothetical decision maker (DM) assessed the values of z_1^0, z_1^1, z_2^0 and z_2^1 to be -40, -60, 200 and 150, respectively. Then the corresponding membership functions $\mu_1(z_1)$ and $\mu_2(z_2)$ become as follows:

$$\mu_1(z_1) = \begin{cases} 1, & z_1 \leq -60 \\ -\dfrac{1}{20}(z_1 + 40), & -60 < z_1 \leq -40 \\ 0, & z_1 > -40, \end{cases} \quad (5.13)$$

$$\mu_2(z_2) = \begin{cases} 1, & z_2 \leq 150 \\ -\dfrac{1}{50}(z_2 - 200), & 150 < z_2 \leq 200 \\ 0, & z_2 > 200. \end{cases} \quad (5.14)$$

The fuzzy goals defined by these membership functions are depicted in Figures 5.2 and 5.3, respectively.

Then the multiobjective linear programming problem incorporating the fuzzy goals is reduced to the linear programming problem with the block angular structure

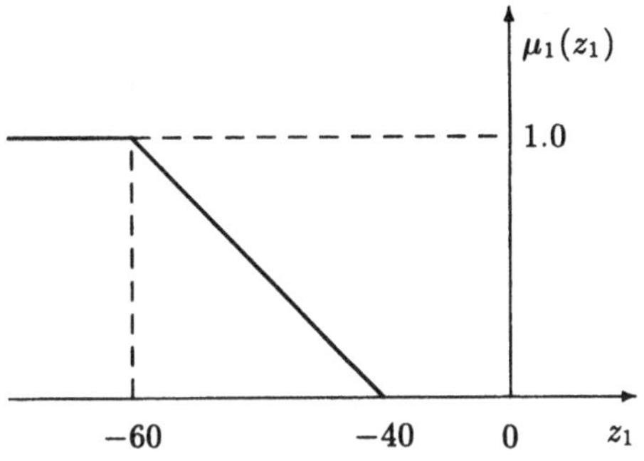

Figure 5.2. Fuzzy goal with respect to z_1.

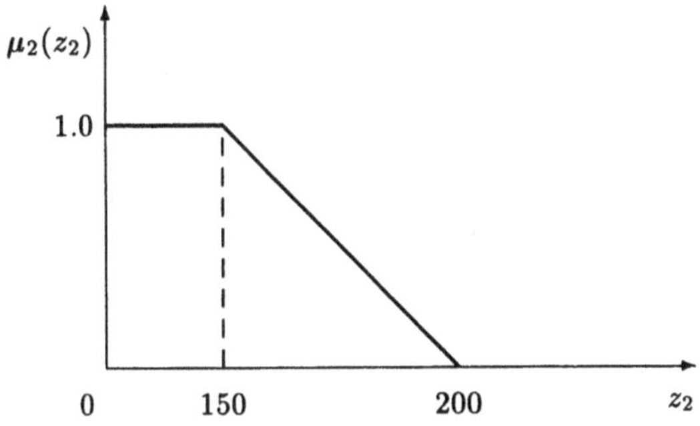

Figure 5.3. Fuzzy goal with respect to z_2.

$$\left.\begin{array}{lllllllll}
\text{maximize} & \lambda \\
\text{subject to} & 0.05w_1 + & 0.1w_2 + 0.05x_1 + 0.05x_2 + & 0.1y_1 + 0.05y_2 - \lambda \geq 2 \\
& 0.06w_1 + 0.16w_2 + 0.08x_1 + 0.06x_2 + 0.12y_1 + 0.08y_2 + \lambda \leq 4 \\
& w_1 + w_2 + x_1 + 2x_2 + 2y_1 + y_2 \leq 60 \\
& 2w_1 + 3w_2 \leq 40 \\
& w_1 + 4w_2 \leq 40 \\
& 5w_1 + w_2 \leq 61 \\
& x_1 + 3x_2 \leq 30 \\
& 2x_1 + x_2 \leq 20 \\
& y_1 \leq 10 \\
& y_2 \leq 10 \\
& y_1 + y_2 \leq 15 \\
& w_i \geq 0,\ x_i \geq 0,\ y_i \geq 0,\ i = 1, 2,
\end{array}\right\} \tag{5.15}$$

to which the Dantzig-Wolfe decomposition method is applicable.

To clarify the solution procedures for solving the problem (5.15) through the Dantzig-Wolfe decomposition method, in the following, we illustrate the first and second iterations.

Iteration 1

Using the phase one of the two phase method, we obtain an initial basic feasible solution $(w_1, w_2, x_1, x_2, y_1, y_2, \lambda) = (11, 6, 6, 8, 10, 0, 0.022)$ with the corresponding simplex multipliers

$$(\pi_{01}, \pi_{02}, \pi_{03}, \bar{\pi}_1, \bar{\pi}_2, \bar{\pi}_3) = (0, -1, 0, 1.62, 0.96, 1.2).$$

Observe that an initial basic feasible solution is obtainable through the Dantzig-Wolfe decomposition method. For notational convenience, let X_1, X_2 or X_3, respectively, denote the feasible region satisfying each of the block constraints of the problem (5.15).

Solving the corresponding three independent subproblems

$$\underset{(w_1,w_2)\in X_1}{\text{minimize}} \ \ 0.06w_1 + 0.16w_2$$

$$\underset{(x_1,x_2)\in X_2}{\text{minimize}} \ \ 0.08x_1 + 0.06x_2$$

$$\underset{(y_1,y_2)\in X_3}{\text{minimize}} \ \ 0.12y_1 + 0.08y_2$$

yields optimal solutions $(w_1, w_2) = (0, 0)$, $(x_1, x_2) = (0, 0)$ and $(y_1, y_2) = (0, 0)$, respectively.

Since $z_1^0(0, -1, 0) - \bar{\pi}_1 = -1.62 < 0$, $z_2^0(0, -1, 0) - \bar{\pi}_2 = -1.96 < 0$ and $z_3^0(0, -1, 0) - \bar{\pi}_3 = -1.20 < 0$, the initial basic feasible solution is not an optimal solution. Thus, solving a linear programming problem

$$\begin{array}{ll}
\text{minimize} & -\lambda \\
\text{subject to} & 1.15\alpha_1 + 0.7\beta_1 + \gamma_1 - \lambda \quad \geq 2 \\
& 1.62\alpha_1 + 0.96\beta_1 + 1.2\gamma_1 + \lambda \leq 4 \\
& 17.0\alpha_1 + 22.0\beta_1 + 20.0\gamma_1 \quad \leq 60 \\
& \alpha_1 + \alpha_2 \quad = 1 \\
& \beta_1 + \beta_2 \quad = 1 \\
& \gamma_1 + \gamma_2 \quad = 1
\end{array}$$

yields an optimal solution

$$(\alpha_1, \beta_1, \gamma_1, \alpha_2, \beta_2, \gamma_2, \lambda) = (0.77256, 1, 1, 0.22744, 0, 0, 0.58845),$$

where β_2 and γ_2 are nonbasic variables. The new simplex multipliers are obtained as

$$(\pi_{01}, \pi_{02}, \pi_{03}, \bar{\pi}_1, \bar{\pi}_2, \bar{\pi}_3) = (0.58484, -0.41516, 0, 0, -0.01083, -0.08664).$$

Iteration 2

Solving three independent subproblems

$$\underset{(w_1,w_2)\in X_1}{\text{minimize}} \ -0.0043324w_1 + 0.0079416w_2$$

$$\underset{(x_1,x_2)\in X_2}{\text{minimize}} \ 0.0039708x_1 + 0.043324x_2$$

$$\underset{(y_1,y_2)\in X_3}{\text{minimize}} \ -0.0086648y_1 + 0.039708y_2$$

yields optimal solutions $(w_1, w_2) = (12.2, 0)$, $(x_1, x_2) = (0, 10)$, $(y_1, y_2) = (10, 0)$, respectively. Since $z_1^0(0.58484, -0.41516, 0) - \bar{\pi}_1 = -0.05286 < 0$, $z_2^0(0.58484, -0.41516, 0) - \bar{\pi}_2 = -0.03249 < 0$ and $z_3^0(0.58484, -0.41516, 0) - \bar{\pi}_3 = -0.00001 < 0$, the current feasible solution is not an optimal solution. Thus, solving a linear programming problem

$$\begin{array}{ll} \text{minimize} & -\lambda \\ \text{subject to} & \\ 1.15\alpha_1 + 0.7\beta_1 + \gamma_1 + 0.61\alpha_3 + 0.5\beta_3 + \gamma_3 - \lambda & \geq 2 \\ 1.62\alpha_1 + 0.96\beta_1 + 1.2\gamma_1 + 0.732\alpha_3 + 0.6\beta_3 + 1.2\gamma_3 + \lambda & \leq 4 \\ 17\alpha_1 + 22\beta_1 + 20\gamma_1 + 12.2\alpha_3 + 20\beta_3 + 20\gamma_3 & \leq 60 \\ \alpha_1 + \alpha_2 + \alpha_3 & = 1 \\ \beta_1 + \beta_3 & = 1 \\ \gamma_1 + \gamma_3 & = 1 \end{array}$$

yields an optimal solution

$$(\alpha_1, \beta_1, \gamma_1, \alpha_2, \alpha_3, \beta_3, \gamma_3, \lambda) = (0.95098, 0, 1, 0, 0.04902, 0, 1, 0, 0.62353),$$

where β_1, α_2 and γ_3 are nonbasic variables. The new simplex multipliers are obtained as

$$(\pi_{01}, \pi_{02}, \pi_{03}, \bar{\pi}_1, \bar{\pi}_2, \bar{\pi}_3)$$
$$= (0.62185, -0.37815, 0, -0.10252, -0.08403, -0.16807).$$

In this way, at the fifth iteration, the following optimal solution is obtained.

$$(w_1, w_2, x_1, x_2, y_1, y_2, \lambda) = (11.605, 2.9748, 0, 10, 10, 5, 0.62773).$$

5.4 Interactive fuzzy multiobjective linear programming

Having elicited the membership functions $\mu_i(\boldsymbol{c}_i\boldsymbol{x})$, $i = 1,\ldots,k$, from the decision maker (DM) for each of the objective functions $\boldsymbol{c}_i\boldsymbol{x}$, $i = 1,\ldots,k$, the original problem (5.1) can be converted into the fuzzy multiobjective optimization problem defined by

$$\underset{\boldsymbol{x}\in X}{\text{maximize}}\ (\mu_1(\boldsymbol{c}_1\boldsymbol{x}),\ldots,\mu_k(\boldsymbol{c}_k\boldsymbol{x})). \tag{5.16}$$

By introducing a general aggregation function

$$\mu_D(\boldsymbol{\mu}(\boldsymbol{c}\boldsymbol{x})) = \mu_D(\mu_1(\boldsymbol{c}_1\boldsymbol{x}),\ldots,\mu_k(\boldsymbol{c}_k\boldsymbol{x})) \tag{5.17}$$

a general fuzzy multiobjective decision making problem can be defined by

$$\underset{\boldsymbol{x}\in X}{\text{maximize}}\ \mu_D(\boldsymbol{\mu}(\boldsymbol{c}\boldsymbol{x})). \tag{5.18}$$

Observe that the value of $\mu_D(\boldsymbol{\mu}(\boldsymbol{c}\boldsymbol{x}))$ can be interpreted as representing an overall degree of satisfaction with the DM's multiple fuzzy goals.

The fuzzy decision or the minimum operator of Bellman and Zadeh [3]

$$\min\,(\mu_1(\boldsymbol{c}_1\boldsymbol{x}),\ldots,\mu_k(\boldsymbol{c}_k\boldsymbol{x})) \tag{5.19}$$

can be viewed only as one special example of $\mu_D(\boldsymbol{\mu}(\boldsymbol{c}\boldsymbol{x}))$.

In the conventional fuzzy approaches, it has been implicitly assumed that the minimum operator is the proper representation of the DM's fuzzy preferences, and hence, (5.18) has been interpreted as

$$\underset{\boldsymbol{x}\in X}{\text{maximize}}\ \min_{i=1,\ldots,k}\ \mu_i(\boldsymbol{c}_i\boldsymbol{x}) \tag{5.20}$$

or equivalently

$$\left.\begin{array}{ll} \text{maximize} & \lambda \\ \text{subject to} & \mu_i(\boldsymbol{c}_i\boldsymbol{x}) \geq \lambda,\ i = 1,\ldots,k \\ & \boldsymbol{x}\in X. \end{array}\right\} \tag{5.21}$$

However, it should be emphasized here that this approach is preferable only when the DM feels that the minimum operator is appropriate. In other words, in general decision situations, the DM does not always use the minimum operator when combining the fuzzy goals and/or constraints. Probably the most crucial problem in (5.18) is the identification of an appropriate aggregation function which well represents the DM's fuzzy preferences. If $\mu_D(\cdot)$

can be explicitly identified, then (5.18) reduces to a standard mathematical programming problem. However, this rarely happens, and as an alternative, an interaction with the DM is necessary for finding the satisficing solution of (5.18). In our interactive fuzzy multiobjective linear programming, to generate a candidate for the satisficing solution which is also Pareto optimal, the DM is asked to specify the aspiration levels of achievement for the membership values of all membership functions, called the reference membership levels. The reference membership levels can be viewed as natural extensions of the reference point of Wierzbicki [135, 136] in objective function spaces.

For the DM's reference membership levels $\bar{\boldsymbol{\mu}} = (\bar{\mu}_1, \ldots, \bar{\mu}_k)^T$, the corresponding Pareto optimal solution, which is nearest to the requirements in the minimax sense or better than that if the reference membership levels are attainable, is obtained by solving the following minimax problem

$$\underset{\boldsymbol{x} \in X}{\text{minimize}} \max_{i=1,\ldots,k} \{\bar{\mu}_i - \mu_i(\boldsymbol{c}_i \boldsymbol{x})\}. \tag{5.22}$$

By introducing the auxiliary variable λ, this problem can be equivalently transformed as

$$\left.\begin{array}{ll} \text{minimize} & \lambda \\ \text{subject to} & \bar{\mu}_i - \mu_i(\boldsymbol{c}_i \boldsymbol{x}) \leq \lambda, \; i = 1, \ldots, k \\ & \boldsymbol{x} \in X. \end{array}\right\} \tag{5.23}$$

Observe that this problem preserves linearity as well as the block angular structure and hence the Dantzig-Wolfe decomposition method is applicable.

The relationships between the optimal solutions of the minimax problem and the Pareto optimal concept of the original problem (5.1) can be characterized by the following theorems.

Theorem 5.4.1. *If $\boldsymbol{x}^*$ is a unique optimal solution to the minimax problem* (5.23) *for some $\bar{\mu}_i$, $i = 1, \ldots, k$, then $\boldsymbol{x}^*$ is a Pareto optimal solution to the original problem* (5.1).

Proof. Assume that $\boldsymbol{x}^*$ is not a Pareto optimal solution to the original problem (5.1). Then there exists $\boldsymbol{x} \in X$ such that $\boldsymbol{c}_i \boldsymbol{x} \leq \boldsymbol{c}_i \boldsymbol{x}^*$, or equivalently, $\bar{\mu}_i - \mu_i(\boldsymbol{c}_i \boldsymbol{x}) \geq \bar{\mu}_i - \mu_i(\boldsymbol{c}_i \boldsymbol{x}^*)$ for $i = 1, \ldots, k$ with strict inequality holding at least one j. Hence, it follows that

$$\max_{i=1,\ldots,k} \{\bar{\mu}_i - \mu_i(\boldsymbol{c}_i \boldsymbol{x})\} > \max_{i=1,\ldots,k} \{\bar{\mu}_i - \mu_i(\boldsymbol{c}_i \boldsymbol{x}^*)\}.$$

which contradicts the fact that $\boldsymbol{x}^*$ is an optimal solution to the minimax problem (5.23). Hence, $\boldsymbol{x}^*$ is a Pareto optimal solution to the original problem (5.1).

Theorem 5.4.2. *If $\boldsymbol{x}^*$ is a Pareto optimal solution to the original problem* (5.1) *with $0 < \mu_i(\mathbf{c}_i\boldsymbol{x}^*) < 1$ holding for all i, then there exists $\bar{\mu}_i$, $i = 1, \ldots, k$, such that $\boldsymbol{x}^*$ is an optimal solution to the minimax problem* (5.23).

Proof. Assume that $\boldsymbol{x}^*$ is not an optimal solution to the minimax problem (5.23) or equivalently (5.22) for any $(\bar{\mu}_1, \ldots, \bar{\mu}_k)^T$ satisfying

$$\bar{\mu}_1 - \mu_1(\mathbf{c}_1\boldsymbol{x}^*) = \cdots = \bar{\mu}_k - \mu_k(\mathbf{c}_k\boldsymbol{x}^*).$$

Then there exists $\boldsymbol{x} \in X$ such that

$$\max_{i=1,\ldots,k} \{\bar{\mu}_i - \mu_i(\mathbf{c}_i\boldsymbol{x}^*)\} > \max_{i=1,\ldots,k} \{\bar{\mu}_i - \mu_i(\mathbf{c}_i\boldsymbol{x})\}.$$

This implies

$$\max_{i=1,\ldots,k} \{\mu_i(\mathbf{c}_i\boldsymbol{x}^*) - \mu_i(\mathbf{c}_i\boldsymbol{x})\} < 0.$$

Now if either any $\mu_i(\mathbf{c}_i\boldsymbol{x}^*) - \mu_i(\mathbf{c}_i\boldsymbol{x})$ is positive or all $\mu_i(\mathbf{c}_i\boldsymbol{x}^*) - \mu_i(\mathbf{c}_i\boldsymbol{x})$, $i = 1, \ldots, k$, are zero, this inequality would be violated. Hence, it follows that

$$\mu_i(\mathbf{c}_i\boldsymbol{x}^*) - \mu_i(\mathbf{c}_i\boldsymbol{x}) \leq 0, \; i = 1, \ldots, k$$

with strict inequality holding for at least one j. Because of the assumption $0 < \mu_i(\mathbf{c}_i\boldsymbol{x}^*) < 1$, we have $\mathbf{c}_i\boldsymbol{x}^* \geq \mathbf{c}_i\boldsymbol{x}$, $i = 1, \ldots, k$ with strict inequality holding for at least one j. This contradicts the fact that $\boldsymbol{x}^*$ is a Pareto optimal solution to the problem (5.1) and the theorem is proved.

From Theorem 5.4.1 and its proof, it can be easily understood that if the uniqueness of the optimal solution $\boldsymbol{x}^*$ to the minimax problem is not guaranteed, only weak Pareto optimality is guaranteed. However, it should be stressed here that the satisficing solution for the DM should be selected from the set of Pareto optimal solutions of the original problem (5.1). As a result, it is necessary to perform the Pareto optimality test of $\boldsymbol{x}^*$.

As discussed in Theorem 5.3.3, the Pareto optimality test for $\boldsymbol{x}^*$ can be performed by solving the following linear programming problem by applying the Dantzig-Wolfe decomposition method.

$$\left.\begin{array}{ll} \text{maximize} & \sum_{i=1}^{k} \varepsilon_i \\ \text{subject to} & \boldsymbol{c}_i \boldsymbol{x} + \varepsilon_i = \boldsymbol{c}_i \boldsymbol{x}^*, \ i = 1, \ldots, k \\ & \boldsymbol{x} \in X \\ & \boldsymbol{\varepsilon} = (\varepsilon_1, \ldots, \varepsilon_k) \geq \mathbf{0} \end{array}\right\} \tag{5.24}$$

For the optimal solutions $(\widehat{\boldsymbol{x}}, \widehat{\boldsymbol{\varepsilon}})$ of this linear programming problem, as discussed in Theorem 5.3.3, (a) If all $\widehat{\varepsilon}_i = 0$, then $\boldsymbol{x}^*$ is a Pareto optimal solution to the original problem (5.1), and (b) If at least one $\widehat{\varepsilon}_i > 0$, not $\boldsymbol{x}^*$ but $\widehat{\boldsymbol{x}}$ is a Pareto optimal solution to the original problem (5.1). Therefore, in case of (b), instead of $\boldsymbol{x}^*$, $\widehat{\boldsymbol{x}}$ which satisfies the Pareto optimality becomes the satisficing solution for the DM incorporating the fuzzy goals.

The DM must either be satisfied with the current Pareto optimal solution or act on this solution by updating the reference membership levels. In order to help the DM express a degree of preference, trade-off information between a standing membership function $\mu_1(\boldsymbol{c}_1\boldsymbol{x})$ and each of the other membership functions is very useful. Such trade-off information is easily obtainable since it is closely related to the simplex multipliers of the minimax problem (5.23).

Let the simplex multipliers corresponding to the constraints $\bar{\mu}_i - \mu_i(\boldsymbol{c}_i\boldsymbol{x}) \leq \lambda$, $i = 1, \ldots, k$ of the minimax problem (5.23) be denoted by $\pi_i^* = \pi_i(\boldsymbol{x}^*)$, $i = 1, \ldots, k$, where $\boldsymbol{x}^*$ is an optimal solution of the minimax problem (5.23). If $\boldsymbol{x}^*$ is a nondegenerate solution of (5.23) and all the constraints of (5.23) are active, then by using the results in Haimes and Chankong [31], the trade-off information between the objective functions can be represented by

$$-\frac{\partial(\boldsymbol{c}_1\boldsymbol{x})}{\partial(\boldsymbol{c}_i\boldsymbol{x})} = -\pi_i^*, \ i = 2, \ldots, k. \tag{5.25}$$

Hence, by the chain rule, the trade-off information between the membership functions is given by

$$-\frac{\partial \mu_1(\boldsymbol{c}_1\boldsymbol{x})}{\partial \mu_i(\boldsymbol{c}_i\boldsymbol{x})} = \frac{\partial \mu_1(\boldsymbol{c}_1\boldsymbol{x})}{\partial(\boldsymbol{c}_1\boldsymbol{x})} \frac{\partial(\boldsymbol{c}_1\boldsymbol{x})}{\partial(\boldsymbol{c}_i\boldsymbol{x})} \left\{ \frac{\partial \mu_i(\boldsymbol{c}_i\boldsymbol{x})}{\partial(\boldsymbol{c}_i\boldsymbol{x})} \right\}^{-1}, \ i = 2, \ldots, k. \tag{5.26}$$

Therefore, for each $i = 2, \ldots, k$, we have the following expression:

$$-\frac{\partial \mu_1(\boldsymbol{c}_1\boldsymbol{x})}{\partial \mu_i(\boldsymbol{c}_i\boldsymbol{x})} = \pi_i^* \frac{\partial \mu_1(\boldsymbol{c}_1\boldsymbol{x})/\partial(\boldsymbol{c}_1\boldsymbol{x})}{\partial \mu_i(\boldsymbol{c}_i\boldsymbol{x})/\partial(\boldsymbol{c}_i\boldsymbol{x})}, \ i = 2, \ldots, k. \tag{5.27}$$

It should be stressed here that in order to obtain the trade-off rate information from (5.27) all the constraints of the minimax problem (5.23) must be active. Therefore, if there are inactive constraints, it is necessary to replace $\bar{\mu}_i$ for inactive constraints by $\mu_i(\boldsymbol{c}_i\boldsymbol{x}^*)$ and solve the corresponding problem to obtain the simplex multipliers.

We can now construct the interactive algorithm in order to derive the satisficing solution for the DM from the Pareto optimal solution set where the steps marked with an asterisk involve interaction with the DM.

Interactive fuzzy multiobjective linear programming

Step 0: Calculate the individual minimum $z_i^{\min}$ and maximum $z_i^{\max}$ of each objective function under the given constraints by solving the problems (5.6) and (5.7) via the Dantzig-Wolfe decomposition method.

Step 1*: By taking account of the calculated individual minimum and maximum, the DM assesses z_i^0 and z_i^1 within $[z_i^{\min}, z_i^{\max}]$ for the determination of the linear membership functions (5.2) for each of the objective functions.

Step 2: Set the initial reference membership levels $\bar{\mu}_i$, $i = 1, \ldots, k$ for the minimax problem (5.23) to 1.

Step 3: For the reference membership levels $\bar{\mu}_i$, $i = 1, \ldots, k$, solve the corresponding minimax problem (5.23) and the Pareto optimality test problem (5.24) by applying the Dantzig-Wolfe decomposition method in order to obtain the Pareto optimal solution and the membership function value. Also obtain the trade-off rate information between the membership functions using the expression (5.27).

Step 4*: If the DM is satisfied with the current levels of the Pareto optimal solution, stop. Then the current Pareto optimal solution is the satisficing solution for the DM. Otherwise, ask the DM to update the current reference membership levels by considering the current values of the membership functions together with the trade-off rates between the membership functions and return to Step 3.

It should be stressed to the DM that any improvement of one membership function can be achieved only at the expense of at least one of the other membership functions.

Example 5.4.1. To illustrate the proposed method, consider the same simple numerical example (5.12) considered in the previous section. Assume that a hypothetical decision maker (DM) assessed the same values for z_1^0, z_1^1, z_2^0 and z_2^1 as in the previous section, by considering the calculated individual minimum and maximum, then the corresponding linear membership functions $\mu_1(z_1)$ and $\mu_2(z_2)$ become same as in (5.13) and (5.14).

Then for the initial reference membership levels $(1,1)^T$, the corresponding minimax problem to be solved becomes as

$$\begin{array}{llllllll}
\text{maximize} & \lambda & & & & & & \\
\text{subject to} & 0.05w_1 + & 0.1w_2 + 0.05x_1 + 0.05x_2 + & 0.1y_1 + 0.05y_2 - \lambda \geq 3 \\
& 0.06w_1 + 0.16w_2 + 0.08x_1 + 0.06x_2 + 0.12y_1 + 0.08y_2 + \lambda \leq 3 \\
& w_1 + w_2 + x_1 + 2x_2 + 2y_1 + y_2 \leq 60 \\
& 2w_1 + 3w_2 \leq 40 \\
& w_1 + 4w_2 \leq 40 \\
& 5w_1 + w_2 \leq 61 \\
& x_1 + 3x_2 \leq 30 \\
& 2x_1 + x_2 \leq 20 \\
& y_1 \leq 10 \\
& y_2 \leq 10 \\
& y_1 + y_2 \leq 15 \\
& w_i \geq 0, \; x_i \geq 0, \; y_i \geq 0, \; i = 1,2.
\end{array}$$

Solving this problem by applying the Dantzig-Wolfe decomposition method yields the Pareto optimal solution

$$z_1 = -52.55, \quad z_2 = 168.61;$$

$$(w_1, w_2, x_1, x_2, y_1, y_2) = (11.605, 2.9748, 0, 10, 10, 5),$$

the membership values

$$\mu_1 = 0.62773, \quad \mu_2 = 0.62773,$$

and the trade-off rates between the membership functions

$$-\frac{\partial \mu_2}{\partial \mu_1} = 1.644.$$

On the basis of such information, the DM updates the reference membership levels to

$$\bar{\mu}_1 = 0.8, \quad \bar{\mu}_2 = 0.5,$$

improving the satisfaction levels for μ_1 at the expense of μ_2.

For the updated reference membership levels, the corresponding minimax problem yields the Pareto optimal solution

$$z_1 = -54.81, \quad z_2 = 177.98;$$

$$(w_1, w_2, x_1, x_2, y_1, y_2) = (11.239, 4.0806, 0, 10, 10, 3.9552),$$

the membership values

$$\mu_1 = 0.7926, \quad \mu_2 = 0.3567,$$

and the trade-off rates

$$-\frac{\partial \mu_2}{\partial \mu_1} = 1.680.$$

If the DM is satisfied with the current values of the membership functions, the procedure stops. Otherwise, a similar procedure continues until the satisficing solution for the DM is obtained.

6. Large Scale Multiobjective Linear Programming with Fuzzy Numbers

In this chapter, in contrast to the large scale multiobjective linear programming problems with block angular structures discussed thus far, by considering the experts' imprecise or fuzzy understanding of the nature of the parameters in the problem-formulation process, large scale multiobjective linear programming problems with block angular structures involving fuzzy numbers are formulated. Through the introduction of extended Pareto optimality concepts, interactive decision making methods, using the Dantzig-Wolfe decomposition method, both without and with the fuzzy goals of the decision maker, for deriving a satisficing solution for the decision maker from the extended Pareto optimal solution set are presented together with detailed numerical examples.

6.1 Introduction

In the previous chapter, we focused mainly on large scale multiobjective linear programming problems with block angular structures and presented fuzzy multiobjective linear programming, and interactive fuzzy multiobjective linear programming to derive the satisficing solution for the decision maker (DM) efficiently from a Pareto optimal solution set.

However, when formulating large scale multiobjective linear programming problems with block angular structures which closely describe and represent the real-world decision situations, various factors of the real-world systems should be reflected in the description of the objective functions and the constraints. Naturally, these objective functions and constraints involve many parameters whose possible values may be assigned by the experts. In the conventional approaches, such parameters are required to fix at some values in an experimental and/or subjective manner through the experts' understanding of the nature of the parameters in the problem-formulation process.

It must be observed here that, in most real-world situations, the possible values of these parameters are often only imprecisely or ambiguously known to the experts. With this observation, it would be certainly more appropriate to interpret the experts' understanding of the parameters as fuzzy numerical data which can be represented by means of fuzzy sets of the real line known as fuzzy numbers [18, 19]. The resulting large scale multiobjective linear programming problems with block angular structures involving fuzzy numbers would be viewed as more realistic versions than the conventional ones.

For dealing with large scale multiobjective linear programming problems with block angular structures involving fuzzy parameters characterized by fuzzy numbers, Sakawa et al. [83, 84, 87, 91] recently introduced the extended Pareto optimality concepts. Then they presented two types of interactive decision making methods, using the Dantzig-Wolfe decomposition method, which are applicable and promising for handling and tackling not only the experts's fuzzy understanding of the nature of parameters in the problem-formulation process but also the fuzzy goals of the DM. These methods can be viewed as a natural generalization of the previous results for multiobjective linear programming problems with block angular structures by Sakawa et al. [80, 81, 102, 103]. In this chapter, these two types of interactive decision making methods for large scale multiobjective linear programming problems with block angular structures involving fuzzy numbers to derive the satisficing solution for the DM efficiently from the extended Pareto optimal solution set are presented in detail together with numerical examples and a lot of computational experiences.

6.2 Interactive multiobjective linear programming with fuzzy numbers

6.2.1 Problem formulation and solution concepts

First, recall the following large scale multiobjective linear programming problem with the block angular structure discussed in Chapter 5:

$$\left.\begin{array}{lrlr}
\text{minimize} & c_1 x = & c_{11}x_1 + \cdots + c_{1p}x_p & \\
& & \vdots & \\
\text{minimize} & c_k x = & c_{k1}x_1 + \cdots + c_{kp}x_p & \\
\text{subject to} & Ax = & A_1x_1 + \cdots + A_px_p & \leq b_0 \\
& & B_1x_1 & \leq b_1 \\
& & \ddots & \vdots \\
& & B_px_p & \leq b_p \\
& & x_j \geq 0, \ j = 1, \ldots, p &
\end{array}\right\} \tag{6.1}$$

where c_{ij}, $i = 1, \ldots, k$, $j = 1, \ldots, p$, are n_j dimensional cost factor row vectors, x_j, $j = 1, \ldots, p$, are n_j dimensional vectors of decision variables, $Ax = A_1x_1 + \cdots + A_px_p \leq b_0$ are m_0 dimensional coupling constraints, and A_j, $j = 1, \ldots, p$, are $m_0 \times n_j$ coefficient matrices. $B_jx_j \leq b_j$, $j = 1, \ldots, p$, are m_j dimensional block constraints with respect to x_j, and B_j, $j = 1, \ldots, p$, are $m_j \times n_j$ coefficient matrices. In the following, for notational convenience, let X denote the feasible region satisfying all of the constraints of the problem (6.1). For simplicity in notation, define the following matrices and vectors.

$$A = (A_1, \ldots, A_p), \quad x = (x_1^T, \ldots, x_p^T)^T \tag{6.2}$$

$$B = \begin{bmatrix} B_1 & & O \\ & \ddots & \\ O & & B_p \end{bmatrix}, \quad b = (b_0^T, b_1^T, \ldots, b_p^T)^T \tag{6.3}$$

$$c_i = (c_{i1}, \ldots, c_{ip}), \ i = 1, \ldots, k, \quad c = (c_1, \ldots, c_k) \tag{6.4}$$

In general, however, for multiobjective linear programming problems, a complete optimal solution which simultaneously minimizes all of the multiple objective functions does not always exist, as the objective functions may conflict with each other. Thus, instead of a complete optimal solution, a Pareto optimal solution defined as Definition 2.4.2 is introduced in the multiobjective linear programming.

In practice, however, it would certainly be more appropriate to consider that the possible values of the parameters in the description of the objective functions and the constraints usually involve the ambiguity of the experts' understanding of the real system. For this reason, in this chapter, we consider the following large scale multiobjective linear programming problem with the block angular structure involving fuzzy parameters [94, 83, 84, 85, 86, 87]:

$$
\left.\begin{array}{lrl}
\text{minimize} & \tilde{\boldsymbol{c}}_1 \boldsymbol{x} = & \tilde{\boldsymbol{c}}_{11}\boldsymbol{x}_1 + \cdots + \tilde{\boldsymbol{c}}_{1p}\boldsymbol{x}_p \\
 & & \vdots \\
\text{minimize} & \tilde{\boldsymbol{c}}_k \boldsymbol{x} = & \tilde{\boldsymbol{c}}_{k1}\boldsymbol{x}_1 + \cdots + \tilde{\boldsymbol{c}}_{kp}\boldsymbol{x}_p \\
\text{subject to} & \tilde{A}\boldsymbol{x} = & \tilde{A}_1\boldsymbol{x}_1 + \cdots + \tilde{A}_p\boldsymbol{x}_p \leq \tilde{\boldsymbol{b}}_0 \\
 & & \tilde{B}_1\boldsymbol{x}_1 \qquad\qquad\quad \leq \tilde{\boldsymbol{b}}_1 \\
 & & \qquad \ddots \qquad\qquad\quad \vdots \\
 & & \qquad\qquad \tilde{B}_p\boldsymbol{x}_p \leq \tilde{\boldsymbol{b}}_p \\
 & & \boldsymbol{x}_j \geq \boldsymbol{0},\ j = 1, \ldots, p
\end{array}\right\} \tag{6.5}
$$

where $\tilde{A}_j$ and $\tilde{B}_j$ or $\tilde{\boldsymbol{b}}_j$ and $\tilde{\boldsymbol{c}}_{ij}$ respectively represent matrices or vectors whose elements are fuzzy parameters. These fuzzy parameters are assumed to be characterized as fuzzy numbers defined by Definition 2.2.1.

Observing that the problem (6.5) involves fuzzy numbers both in the objective functions and the constraints, it is evident that the notion of Pareto optimality defined for the problem (6.1) cannot be applied directly. Thus, it seems essential to extend the notion of usual Pareto optimality in some sense. For that purpose, the α-level set of all of the vectors and matrices whose elements are fuzzy numbers is first introduced.

Definition 6.2.1 (α-level set).

The α-level set of $(\tilde{A}, \tilde{B}, \tilde{\boldsymbol{b}}, \tilde{\boldsymbol{c}})$ is defined as the ordinary set $(\tilde{A}, \tilde{B}, \tilde{\boldsymbol{b}}, \tilde{\boldsymbol{c}})_\alpha$ for which the degree of their membership functions exceeds the level α:

$$
\begin{aligned}
&(\tilde{A}, \tilde{B}, \tilde{\boldsymbol{b}}, \tilde{\boldsymbol{c}})_\alpha \\
&= \{(A, B, \boldsymbol{b}, \boldsymbol{c}) \mid \mu_{\tilde{a}_{ij}^s}(a_{ij}^s) \geq \alpha, \mu_{\tilde{b}_{hj}^s}(b_{hj}^s) \geq \alpha, \mu_{\tilde{b}_{h'}^{\prime s'}}(b_{h'}^{\prime s'}) \geq \alpha, \mu_{\tilde{c}_j^{ts}}(c_j^{ts}) \geq \alpha, \\
&\qquad i = 1, \ldots, m_0, j = 1, \ldots, n_s, s = 1, \ldots, p, h = 1, \ldots, m_s, \\
&\qquad\quad h' = 1, \ldots, m_{s'}, s' = 0, \ldots, p, t = 1, \ldots, k\}
\end{aligned}
$$

Now suppose that the decision maker (DM) decides that the degree of all of the membership functions of the fuzzy numbers involved in the problem (6.5) should be greater than or equal to some value α. Then for such a degree α, the problem (6.5) can be interpreted as a nonfuzzy multiobjective linear programming problem which depends on the coefficient vector $(A, B, \boldsymbol{b}, \boldsymbol{c}) \in (\tilde{A}, \tilde{B}, \tilde{\boldsymbol{b}}, \tilde{\boldsymbol{c}})_\alpha$. Observe that there exists an infinite number of such nonfuzzy multiobjective linear programming problems depending on the coefficient vector $(A, B, \boldsymbol{b}, \boldsymbol{c}) \in (\tilde{A}, \tilde{B}, \tilde{\boldsymbol{b}}, \tilde{\boldsymbol{c}})_\alpha$, and the values of $(A, B, \boldsymbol{b}, \boldsymbol{c})$ are arbitrary for any $(A, B, \boldsymbol{b}, \boldsymbol{c}) \in (\tilde{A}, \tilde{B}, \tilde{\boldsymbol{b}}, \tilde{\boldsymbol{c}})_\alpha$ in the sense that the degree of all of the membership functions for the fuzzy numbers in the problem (6.5) exceeds the level α. However, if possible, it would be desirable for the DM to choose $(A, B, \boldsymbol{b}, \boldsymbol{c}) \in (\tilde{A}, \tilde{B}, \tilde{\boldsymbol{b}}, \tilde{\boldsymbol{c}})_\alpha$ to minimize the objective functions under the constraints. From such a point of view, for a certain degree α, it seems to be quite natural to have the problem (6.5) as the following nonfuzzy α-multiobjective linear programming problem:

$$\left.\begin{array}{lrl} \text{minimize} & \boldsymbol{c}_1\boldsymbol{x} = & \boldsymbol{c}_{11}\boldsymbol{x}_1 + \cdots + \boldsymbol{c}_{1p}\boldsymbol{x}_p \\ & & \vdots \\ \text{minimize} & \boldsymbol{c}_k\boldsymbol{x} = & \boldsymbol{c}_{k1}\boldsymbol{x}_1 + \cdots + \boldsymbol{c}_{kp}\boldsymbol{x}_p \\ \text{subject to} & A\boldsymbol{x} = & A_1\boldsymbol{x}_1 + \cdots + A_p\boldsymbol{x}_p \leq \boldsymbol{b}_0 \\ & & B_1\boldsymbol{x}_1 \qquad\qquad\qquad \leq \boldsymbol{b}_1 \\ & & \qquad\qquad \ddots \qquad\qquad \vdots \\ & & \qquad\qquad\qquad B_p\boldsymbol{x}_p \leq \boldsymbol{b}_p \\ & & \boldsymbol{x}_j \geq \boldsymbol{0}, \ j = 1, \ldots, p \\ & & (A, B, \boldsymbol{b}, \boldsymbol{c}) \in (\tilde{A}, \tilde{B}, \tilde{\boldsymbol{b}}, \tilde{\boldsymbol{c}})_\alpha \end{array}\right\} \tag{6.6}$$

It should be emphasized here that, in the problem (6.6), $(A, B, \boldsymbol{b}, \boldsymbol{c})$ are treated as decision variables rather than coefficients.

On the basis of the α-level sets of the fuzzy numbers, we can introduce the concept of an α-Pareto optimal solution to the problem (6.6) as a natural extension of the Pareto optimality concept for the problem (6.1), where for simplicity, the feasible region of the problem (6.6) is denoted by $X(A, B, \boldsymbol{b})$.

Definition 6.2.2 (α-Pareto optimal solution).
$\boldsymbol{x}^* \in X(A^*, B^*, \boldsymbol{b}^*)$ for $(A^*, B^*, \boldsymbol{b}^*, \boldsymbol{c}^*) \in (\tilde{A}, \tilde{B}, \tilde{\boldsymbol{b}}, \tilde{\boldsymbol{c}})_\alpha$ is said to be an α-Pareto optimal solution to the problem (6.6) if and only if there does not exist another $\boldsymbol{x} \in X(A, B, \boldsymbol{b})$, $(A, B, \boldsymbol{b}, \boldsymbol{c}) \in (\tilde{A}, \tilde{B}, \tilde{\boldsymbol{b}}, \tilde{\boldsymbol{c}})_\alpha$ such that $\boldsymbol{c}_i\boldsymbol{x} \leq \boldsymbol{c}_i^*\boldsymbol{x}^*$, $i = 1, \ldots, k$, with strict inequality holding for at least one i, where the corresponding values of parameters $(A^*, B^*, \boldsymbol{b}^*, \boldsymbol{c}^*) \in (\tilde{A}, \tilde{B}, \tilde{\boldsymbol{b}}, \tilde{\boldsymbol{c}})_\alpha$ are called α-level optimal parameters.

Observe that α-Pareto optimal solutions and α-level optimal parameters can be obtained through a direct application of the usual scalarizing methods for generating Pareto optimal solutions by regarding the decision variables in the the problem (6.6) as $(\boldsymbol{x}, A, B, \boldsymbol{b}, \boldsymbol{c})$.

However, as can be immediately understood from Definition 6.2.2, in general, α-Pareto optimal solutions to the problem (6.6) consist of an infinite number of points and some kinds of subjective judgments should be added by the DM. Namely, the DM must select a compromise or satisficing solution from the set of α-Pareto optimal solutions based on a subjective value judgment. Therefore, in the following, we present an interactive programming approach to the problem (6.5).

6.2.2 Minimax problems

To generate a candidate for the satisficing solution which is also α-Pareto optimal, in our interactive decision making method, the DM is asked to specify the degree α of the α-level set and the reference levels of achievement of the objective functions. Observe that the idea of the reference levels or the reference point first appeared in Wierzbicki [135, 136]. To be more explicit, for the degree α and reference levels $\bar{z}_i$, $i = 1, \ldots, k$, the corresponding α-Pareto optimal solution, which is nearest to the requirement in the minimax sense, or better than that if the reference levels are attainable, is obtained by solving the following minimax problem in an objective function space:

$$\left.\begin{array}{ll} \text{minimize} & \displaystyle\max_{i=1,\ldots,k} (\boldsymbol{c}_i\boldsymbol{x} - \bar{z}_i) \\ \text{subject to} & \boldsymbol{x} \in X(A, B, \boldsymbol{b}) \\ & (A, B, \boldsymbol{b}, \boldsymbol{c}) \in (\tilde{A}, \tilde{B}, \tilde{\boldsymbol{b}}, \tilde{\boldsymbol{c}})_\alpha \end{array}\right\} \tag{6.7}$$

or equivalently

$$\left.\begin{array}{ll} \text{minimize} & v \\ \text{subject to} & \boldsymbol{c}_i\boldsymbol{x} - \bar{z}_i \leq v, \;\; i = 1, \ldots, k \\ & \boldsymbol{x} \in X(A, B, \boldsymbol{b}) \\ & (A, B, \boldsymbol{b}, \boldsymbol{c}) \in (\tilde{A}, \tilde{B}, \tilde{\boldsymbol{b}}, \tilde{\boldsymbol{c}})_\alpha, \end{array}\right\} \tag{6.8}$$

where $X(A, B, \boldsymbol{b})$ denotes the feasible region satisfying the constraints of the problem (6.6).

In this formulation, however, constraints are nonlinear because the parameters A, B, $\boldsymbol{b}$ and $\boldsymbol{c}$ are treated as decision variables. To deal with such nonlinearities, we introduce the following set-valued functions $S_i(\cdot)$, $T_j(\cdot,\cdot)$ and $U(\cdot,\cdot)$:

$$\left.\begin{array}{rcl} S_i(\boldsymbol{c}_i) & = & \{(\boldsymbol{x}, v) \mid \boldsymbol{c}_i\boldsymbol{x} - \bar{z}_i \leq v, \; i = 1, \ldots, k, \; \boldsymbol{x} \geq \boldsymbol{0}\} \\ T_j(B_j, \boldsymbol{b}_j) & = & \{\boldsymbol{x}_j \mid B_j\boldsymbol{x}_j \leq \boldsymbol{b}_j, \; \boldsymbol{x}_j \geq \boldsymbol{0}\} \\ U(A, \boldsymbol{b}_0) & = & \{\boldsymbol{x} \mid A\boldsymbol{x} \leq \boldsymbol{b}_0, \; \boldsymbol{x} \geq \boldsymbol{0}\}. \end{array}\right\} \tag{6.9}$$

Then it can be easily verified that the following relations hold for $S_i(\cdot)$, $T_j(\cdot,\cdot)$ and $U(\cdot,\cdot)$ when $\boldsymbol{x} \geq \boldsymbol{0}$.

Proposition 6.2.1.

(1) *If* $\boldsymbol{c}_i^1 \leq \boldsymbol{c}_i^2$, *then* $S_i(\boldsymbol{c}_i^1) \supseteq S_i(\boldsymbol{c}_i^2)$.

(2) *If* $B_j^1 \leq B_j^2$, *then* $T_j(B_j^1, \boldsymbol{b}_j) \supseteq T_j(B_j^2, \boldsymbol{b}_j)$.

(3) *If* $\boldsymbol{b}_j^1 \leq \boldsymbol{b}_j^2$, *then* $T_j(B_j, \boldsymbol{b}_j^1) \subseteq T_j(B_j, \boldsymbol{b}_j^2)$.

(4) *If* $A^1 \leq A^2$, *then* $U(A^1, \boldsymbol{b}_0) \supseteq U(A^2, \boldsymbol{b}_0)$.

(5) *If* $\boldsymbol{b}_0^1 \leq \boldsymbol{b}_0^2$, *then* $U(A, \boldsymbol{b}^1) \subseteq U(A, \boldsymbol{b}^2)$.

Now from the properties of the α-level set for the vectors and/or matrices of fuzzy numbers, it should be noted here that the feasible regions for $\boldsymbol{c}_i$, A, B_j and $\boldsymbol{b}_j$ can be denoted respectively by the closed intervals $[\boldsymbol{c}_{i\alpha}^L, \boldsymbol{c}_{i\alpha}^R]$, $[A_\alpha^L, A_\alpha^R]$, $[B_{j\alpha}^L, B_{j\alpha}^R]$ and $[\boldsymbol{b}_{j\alpha}^L, \boldsymbol{b}_{j\alpha}^R]$, where y_α^L or y_α^R represents the left or right extreme point of the α-level set $\tilde{y}_\alpha$.

Therefore, through the use of Proposition 6.2.1, we can obtain an optimal solution of problem (6.7) by solving the following linear programming problem:

$$\left.\begin{array}{rlllll}
\text{minimize} & v & & & & \\
\text{subject to} & \boldsymbol{c}_{11\alpha}^L \boldsymbol{x}_1 & + \cdots + & \boldsymbol{c}_{1p\alpha}^L \boldsymbol{x}_p & \leq & v + \bar{z}_1 \\
 & \vdots & & & & \vdots \\
 & \boldsymbol{c}_{k1\alpha}^L \boldsymbol{x}_1 & + \cdots + & \boldsymbol{c}_{kp\alpha}^L \boldsymbol{x}_p & \leq & v + \bar{z}_k \\
 & A_{1\alpha}^L \boldsymbol{x}_1 & + \cdots + & A_{p\alpha}^L \boldsymbol{x}_p & \leq & \boldsymbol{b}_{0\alpha}^R \\
 & B_{1\alpha}^L \boldsymbol{x}_1 & & & \leq & \boldsymbol{b}_{1\alpha}^R \\
 & & \ddots & & & \vdots \\
 & & & B_{p\alpha}^L \boldsymbol{x}_p & \leq & \boldsymbol{b}_{p\alpha}^R \\
 & & \multicolumn{3}{l}{\boldsymbol{x}_j \geq \boldsymbol{0}, \; j = 1, \ldots, p.} &
\end{array}\right\} \quad (6.10)$$

It is important to realize here that this problem is no more nonlinear but is an ordinary linear programming problem. Moreover, in addition to linearity, this problem preserves the block angular structure and hence the Dantzig-Wolfe decomposition method is applicable.

The relationships between the optimal solutions to (6.10) and the α-Pareto optimal concept of the problem (6.6) can be characterized by the following theorems [78, 87].

Theorem 6.2.1. *If* $\boldsymbol{x}^*$ *is a unique optimal solution to* (6.10) *for certain reference points* $\bar{z} = (\bar{z}_1, \ldots, \bar{z}_k)$, *then* $\boldsymbol{x}^*$ *is an* α*-Pareto optimal solution to* (6.6).

Theorem 6.2.2. *If $\boldsymbol{x}^*$ is an α-Pareto optimal solution and $(A^*, B^*, \boldsymbol{b}^*, \boldsymbol{c}^*)$ are α-level optimal parameters to* (6.6), *then $\boldsymbol{x}^*$ is an optimal solution to* (6.10) *for some $\bar{\boldsymbol{z}} = (\bar{z}_1, \dots, \bar{z}_k)$.*

It should be noted here that, to generate α-Pareto optimal solutions using Theorem 6.2.1, the uniqueness of the solution must be verified. In general, however, it is not easy to check numerically whether an optimal solution to (6.10) is unique or not. Consequently, in order to perform a numerical test of α-Pareto optimality of a current optimal solution $\boldsymbol{x}^*$, we formulate and solve the following linear programming problem with decision variables $\boldsymbol{x}$ and ε_i:

$$\left.\begin{array}{ll} \text{maximize} & \varepsilon = \sum_{i=1}^{k} \varepsilon_i \\ \text{subject to} & \boldsymbol{c}_{i\alpha}^L \boldsymbol{x} + \varepsilon_i \ = \ \boldsymbol{c}_{i\alpha}^L \boldsymbol{x}^*, \ i = 1, \dots, k \\ & A_\alpha^L \boldsymbol{x} \ \leq \ \boldsymbol{b}_{0\alpha}^R \\ & B_{j\alpha}^L \boldsymbol{x}_j \ \leq \ \boldsymbol{b}_{j\alpha}^R, \ j = 1, \dots, p \\ & \boldsymbol{x}_j \geq \boldsymbol{0}, \ j = 1, \dots, p \\ & \varepsilon_i \geq 0, \ i = 1, \dots, k. \end{array}\right\} \quad (6.11)$$

As discussed in Chapter 5, for an optimal solution $(\bar{\boldsymbol{x}}, \bar{\varepsilon})$ to this problem, the following theorems hold:

Theorem 6.2.3. *Let $(\bar{\boldsymbol{x}}, \bar{\boldsymbol{\varepsilon}})$ be an optimal solution to problem* (6.11).
(1) *If $\bar{\boldsymbol{\varepsilon}} = \boldsymbol{0}$, then $\boldsymbol{x}^*$ is an α-Pareto optimal solution to* (6.6).
(2) *If $\bar{\boldsymbol{\varepsilon}} \neq \boldsymbol{0}$, then not $\boldsymbol{x}^*$ but $\bar{\boldsymbol{x}}$ is an α-Pareto optimal solution to* (6.6).

It should be noted here that this problem (6.11) preserves the block angular structure and hence the Dantzig-Wolfe decomposition method is applicable.

6.2.3 Interactive programming

Now, given the α-Pareto optimal solution for the degree α and the reference levels specified by the DM by solving the corresponding minimax problem, the DM must either be satisfied with the current solution or act on this solution by updating the reference levels and/or the degree α. To help the DM express a degree of preference, trade-off information between a standing objective function and each of the other objective functions as well as between the degree α and the objective functions is very useful. Such trade-off information is easily obtainable since it is closely related to the simplex multipliers of the problem (6.10).

To derive the trade-off information, first define the following Lagrangian function L for the problem (6.10) with respect to the coupling constraints:

$$L = v + \sum_{i=1}^{k} \pi_i (\boldsymbol{c}_{i\alpha}^L \boldsymbol{x} - \bar{z}_i - v) + \boldsymbol{\lambda}(A_\alpha^L \boldsymbol{x} - \boldsymbol{b}_{0\alpha}^R). \tag{6.12}$$

If all $\pi_i > 0$ for each i, then by extending the results in Haimes and Chankong [31], it can be proved that the following expression holds:

$$-\frac{\partial(\boldsymbol{c}_{1\alpha}^L \boldsymbol{x})}{\partial(\boldsymbol{c}_{i\alpha}^L \boldsymbol{x})} = \frac{\pi_i}{\pi_1}, \quad i = 2, \ldots, k. \tag{6.13}$$

Regarding a trade-off rate between $\boldsymbol{c}_i \boldsymbol{x}$ and α, the following relation from the sensitivity theorem [21]:

$$\frac{\partial(\boldsymbol{c}_{i\alpha}^L \boldsymbol{x})}{\partial \alpha} = \sum_{i=1}^{k} \pi_i \frac{\partial(\boldsymbol{c}_{i\alpha}^L)}{\partial \alpha} \boldsymbol{x} + \boldsymbol{\lambda} \left\{ \frac{\partial(A_\alpha^L)}{\partial \alpha} \boldsymbol{x} - \frac{\partial(\boldsymbol{b}_{0\alpha}^R)}{\partial \alpha} \right\}, \quad i = 1, \ldots, k. \tag{6.14}$$

It should be noted here that to obtain the trade-off information from (6.13) or (6.14), all the constraints $\boldsymbol{c}_{i\alpha}^L \boldsymbol{x} - \bar{z}_i \leq v$, $i = 1, \ldots, k$, of the problem (6.10) must be active for the current optimal solution. Therefore, if there are inactive constraints, it is necessary to replace $\bar{z}_i$ for the inactive constraints by $\boldsymbol{c}_{i\alpha}^L \boldsymbol{x}^* - v^*$ and solve the corresponding minimax problem (6.10) for obtaining the simplex multipliers.

We can now construct the interactive algorithm to derive the satisficing solution for the DM from the α-Pareto optimal solution set. The steps marked with an asterisk involve interaction with the DM.

Interactive multiobjective linear programming with fuzzy numbers

Step 0: Calculate the individual minimum and maximum of each objective function of the problem (6.10) under the given constraints for $\alpha = 0$ and $\alpha = 1$ via the Dantzig-Wolfe decomposition method.

Step 1*: Ask the DM to select the initial value of α ($0 \leq \alpha \leq 1$) and the initial reference levels $\bar{z}_i$, $i = 1, \ldots, k$.

Step 2: For the degree α and the reference levels $\bar{z}_i$ specified by the DM, using the Dantzig-Wolfe decomposition method, solve the corresponding minimax problem and perform the α-Pareto optimality test to obtain the α-Pareto optimal solution together with the trade-off rates between the objective functions and the degree α.

Step 3*: The DM is supplied with the corresponding α-Pareto optimal solution and the trade-off rates between the objective functions and the degree α. If the DM is satisfied with the current objective function values of the α-Pareto optimal solution, stop. Otherwise, the DM must update the reference levels and/or the degree α by considering the current values of objective functions and α together with the trade-off rates between the objective functions, and the degree α and return to Step 2.

Here it should be stressed to the DM that (1) any improvement of one objective function can be achieved only at the expense of at least one of the other objective functions for some fixed degree α and (2) the greater value of the degree α gives the worse values of the objective functions for some fixed reference levels.

6.2.4 Numerical example

To illustrate the proposed method, consider the following three-objective linear programming problem with fuzzy numbers (20 variables, 30 constraints):

$$\left.\begin{array}{rlllllllll}
\text{minimize} & \tilde{\boldsymbol{c}}_{11}\boldsymbol{x}_1 & + & \tilde{\boldsymbol{c}}_{12}\boldsymbol{x}_2 & + & \tilde{\boldsymbol{c}}_{13}\boldsymbol{x}_3 & + & \tilde{\boldsymbol{c}}_{14}\boldsymbol{x}_4 & & \\
\text{minimize} & \tilde{\boldsymbol{c}}_{21}\boldsymbol{x}_1 & + & \tilde{\boldsymbol{c}}_{22}\boldsymbol{x}_2 & + & \tilde{\boldsymbol{c}}_{23}\boldsymbol{x}_3 & + & \tilde{\boldsymbol{c}}_{24}\boldsymbol{x}_4 & & \\
\text{minimize} & \tilde{\boldsymbol{c}}_{31}\boldsymbol{x}_1 & + & \tilde{\boldsymbol{c}}_{32}\boldsymbol{x}_2 & + & \tilde{\boldsymbol{c}}_{33}\boldsymbol{x}_3 & + & \tilde{\boldsymbol{c}}_{34}\boldsymbol{x}_4 & & \\
\text{subject to} & \tilde{A}_1\boldsymbol{x}_1 & + & \tilde{A}_2\boldsymbol{x}_2 & + & \tilde{A}_3\boldsymbol{x}_3 & + & \tilde{A}_4\boldsymbol{x}_4 & [\leq \text{ or } \geq] & \tilde{\boldsymbol{b}}_0 \\
 & B_1\boldsymbol{x}_1 & & & & & & & & \leq \boldsymbol{b}_1 \\
 & & & B_2\boldsymbol{x}_2 & & & & & & \leq \boldsymbol{b}_2 \\
 & & & & & B_3\boldsymbol{x}_3 & & & & \leq \boldsymbol{b}_3 \\
 & & & & & & & B_4\boldsymbol{x}_4 & & \leq \boldsymbol{b}_4 \\
 & & & & & \boldsymbol{x}_i \geq \boldsymbol{0},\ i = 1,\ldots,4 & & & &
\end{array}\right\}$$

where

$$\boldsymbol{x}_1 = (x_1, x_2, x_3, x_4, x_5),\quad \boldsymbol{x}_2 = (x_6, x_7, x_8, x_9, x_{10}),$$

$$\boldsymbol{x}_3 = (x_{11}, x_{12}, x_{13}, x_{14}, x_{15}),\quad \boldsymbol{x}_4 = (x_{16}, x_{17}, x_{18}, x_{19}, x_{20}),$$

$$\tilde{\boldsymbol{c}}_{11} = (-2, 1.2, c_1^1, -4.8, c_1^2),\quad \tilde{\boldsymbol{c}}_{12} = (1, 2, c_1^3, -1.3, -1),$$

$$\tilde{\boldsymbol{c}}_{13} = (c_1^4, -3.5, 1, c_1^5, -9.6),\quad \tilde{\boldsymbol{c}}_{14} = (c_1^6, 1, 1.1, -2.2, 1),$$

$$\tilde{\boldsymbol{c}}_{21} = (7, c_2^1, 3.8, c_2^2, -7.1),\quad \tilde{\boldsymbol{c}}_{22} = (c_2^3, -1.1, 1, -2.3, c_2^4),$$

$$\tilde{\boldsymbol{c}}_{23} = (-1, c_2^5, 3.8, 1, 1),\quad \tilde{\boldsymbol{c}}_{24} = (-2.7, 1.1, c_2^6, 1, c_2^7),$$

$$\tilde{\mathbf{c}}_{31} = (c_3^1, 0, c_3^2, 0, c_3^3), \quad \tilde{\mathbf{c}}_{32} = (0, c_3^4, 0, c_3^5, 0),$$

$$\tilde{\mathbf{c}}_{33} = (c_3^6, 0, c_3^7, 0, c_3^8), \quad \tilde{\mathbf{c}}_{34} = (0, c_3^9, 0, c_3^{10}, 0),$$

$$\tilde{A}_1 = \begin{pmatrix} 1 & 0 & a_1^1 & 0 & a_1^2 \\ 0 & 1 & 0 & a_1^3 & 0 \\ a_1^4 & a_1^5 & 2 & 2 & 3 \\ 1 & 1 & 1 & 1 & 1 \\ 0 & 0 & 0 & 0 & 0 \\ 0 & 0 & 0 & 0 & 0 \\ a_1^6 & 3 & a_1^7 & 1 & 1 \\ a_1^8 & 3 & a_1^9 & 1 & 1 \\ 0 & 0 & 0 & 0 & 0 \\ 0 & 0 & 0 & 0 & 0 \end{pmatrix}, \quad \tilde{A}_2 = \begin{pmatrix} 0 & 1 & 0 & 1 & 0 \\ a_2^1 & 0 & 1 & 0 & 1 \\ 1 & 1 & a_2^2 & 1 & 1 \\ 2 & a_2^3 & 2 & a_2^4 & a_2^5 \\ 0 & 0 & 0 & 0 & 0 \\ 0 & 0 & 0 & 0 & 0 \\ 0 & 0 & 0 & 0 & 0 \\ 0 & 0 & 0 & 0 & 0 \\ 1 & a_2^6 & 3 & a_2^7 & 1 \\ 1 & a_2^8 & 3 & a_2^9 & 1 \end{pmatrix},$$

$$\tilde{A}_3 = \begin{pmatrix} a_3^1 & 0 & a_3^2 & 0 & 1 \\ 0 & 2 & 0 & 4 & 0 \\ 0 & 0 & 0 & 0 & 0 \\ 0 & 0 & 0 & 0 & 0 \\ 2 & a_3^3 & a_3^4 & 1 & a_3^5 \\ a_3^6 & 2.5 & 1 & a_3^7 & 1 \\ 1 & a_3^8 & a_3^9 & 1 & 1 \\ 1 & a_3^{10} & a_3^{11} & 1 & 1 \\ 0 & 0 & 0 & 0 & 0 \\ 0 & 0 & 0 & 0 & 0 \end{pmatrix}, \quad \tilde{A}_4 = \begin{pmatrix} 0 & a_4^1 & 0 & 1 & 0 \\ 3 & 0 & 1 & 0 & a_4^2 \\ 0 & 0 & 0 & 0 & 0 \\ 0 & 0 & 0 & 0 & 0 \\ a_4^3 & 1 & 1 & 1 & 1 \\ 1 & 3 & a_4^4 & 1 & a_4^5 \\ 0 & 0 & 0 & 0 & 0 \\ 0 & 0 & 0 & 0 & 0 \\ a_4^6 & 2 & a_4^7 & 2 & 1 \\ a_4^7 & 2 & a_4^8 & 2 & 1 \end{pmatrix},$$

$$B_1 = \begin{pmatrix} 1 & 2 & 3 & 1 & 1 \\ -1 & 2 & -3 & 1 & 1 \\ 2 & 3 & 1 & 2 & 1 \\ -2 & 2 & 1 & 1 & 3 \\ 3 & 1 & 2 & 1 & 1 \end{pmatrix}, \quad B_2 = \begin{pmatrix} 3 & 1 & 2 & 3 & 1 \\ -3 & 1 & 2 & 3 & -1 \\ 1 & 2 & 1 & 1 & -1 \\ 2 & 1 & -3 & -1 & -1 \\ 1 & -1 & 1 & -1 & 1 \end{pmatrix},$$

$$B_3 = \begin{pmatrix} 1 & -1 & 2 & 1 & 1 \\ 1 & 1 & 2 & -1 & 1 \\ -1 & -1 & 2 & 1 & 1 \\ 2 & 1 & 1 & 2 & 2 \\ 1 & -1 & 1 & -1 & 3 \end{pmatrix}, \quad B_4 = \begin{pmatrix} 1 & 2 & 1 & 3 & 1.5 \\ 1 & 1 & 1.5 & 2 & 1 \\ 1.5 & 2 & 1 & 2.5 & -1 \\ -2 & 1 & 2 & 1 & 1 \\ 1 & 1 & 1 & 1 & 2 \end{pmatrix},$$

$$\tilde{\mathbf{b}}_0 = \begin{pmatrix} 3500 \\ b_0^1 \\ 6000 \\ 4000 \\ b_0^2 \\ 3400 \\ b_0^3 \\ b_0^4 \\ b_0^5 \end{pmatrix}, \quad [\leq \textit{ or } \geq] = \begin{pmatrix} \leq \\ \leq \\ \leq \\ \leq \\ \leq \\ \leq \\ \leq \\ \geq \\ \leq \\ \geq \end{pmatrix},$$

$$\boldsymbol{b}_1 = \begin{pmatrix} 3000 \\ 1000 \\ 4500 \\ 2500 \\ 3300 \end{pmatrix}, \ \boldsymbol{b}_2 = \begin{pmatrix} 2500 \\ 2000 \\ 2200 \\ 1000 \\ 1000 \end{pmatrix}, \ \boldsymbol{b}_3 = \begin{pmatrix} 1800 \\ 1800 \\ 1500 \\ 2300 \\ 2000 \end{pmatrix}, \ \boldsymbol{b}_4 = \begin{pmatrix} 1500 \\ 1500 \\ 2000 \\ 1800 \\ 1000 \end{pmatrix}$$

Here, for simplicity, it is assumed that all of the membership functions for the fuzzy numbers involved in this example are triangular ones and the corresponding parameter values are given in Table 6.1.

Table 6.1. Fuzzy numbers for example

	l	m	r		l	m	r		l	m	r
a_1^1	0.7	1	1.2	a_1^2	−2.3	−2	−1.7	a_1^3	5.2	6	6.8
a_1^4	2.8	3	3.2	a_1^5	1.7	2	2.3	a_1^6	1.6	2	2.4
a_1^7	1.1	1.5	1.7	a_1^8	1.6	2	2.4	a_1^9	1.1	1.5	1.7
a_2^1	1.5	2	2.5	a_2^2	0.8	1	1.3	a_2^3	1.7	2	2.3
a_2^4	2.7	3	3.3	a_2^5	1.6	2	2.5	a_2^6	1.3	2	2.7
a_2^7	2.3	3	3.7	a_2^8	1.3	2	2.7	a_2^9	2.3	3	3.7
a_3^1	2.5	3	3.4	a_3^2	2	2.5	3	a_3^3	2.2	3	3.5
a_3^4	1.2	1.5	1.8	a_3^5	0.8	1	1.2	a_3^6	1	1.2	1.4
a_3^7	0.8	1	1.3	a_3^8	5.5	6	6.5	a_3^9	4.8	5	5.2
a_3^{10}	5.5	6	6.5	a_3^{11}	4.8	5	5.2	a_4^1	1.2	1.5	1.7
a_4^2	1.2	2	2.7	a_4^3	1.7	2	2.4	a_4^4	1.7	2	2.4
a_4^5	0.7	1	1.3	a_4^6	1.2	1.5	1.7	a_4^7	1	1.2	1.5
a_4^8	1.2	1.5	1.8	a_4^9	1	1.2	1.5	b_0^1	3300	3500	3800
b_0^2	2700	3000	3400	b_0^3	3100	3500	3800	b_0^4	850	1000	1200
b_0^5	2700	2800	2900	b_0^6	700	800	900	c_1^1	7	7.3	7.5
c_1^2	−3.6	−3.3	−3	c_1^3	−7	−6.5	−6	c_1^4	7.5	8.2	8.7
c_1^5	−1.5	−1	−0.5	c_1^6	0.1	0.3	0.5	c_2^1	8.3	9	9.8
c_2^2	−5	−4.4	−4	c_2^3	0.8	1.1	1.5	c_2^4	0.7	1	1.3
c_2^5	3.5	3.8	4.3	c_2^6	1	1.3	1.5	c_2^7	−7.5	−7.1	−6.5
c_3^1	2.5	3	3.5	c_3^2	−4.5	−4.1	−3.6	c_3^3	1.3	1.5	1.7
c_3^4	−5	−4.5	−4.2	c_3^5	3.3	3.7	4	c_3^6	−1.5	−1.1	−0.7
c_3^7	0.8	1	1.2	c_3^8	2.2	2.7	3	c_3^9	2.7	3.1	3.5
c_3^{10}	−10	−9.3	−8.5								

First calculating the individual minimum and maximum of each objective functions for $\alpha = 0$ and 1 via the Dantzig-Wolfe decomposition method yields the results shown in Table 6.2.

Now, for illustrative purposes, by considering the individual minimum and maximum for $\alpha = 0$ and 1, assume that the hypothetical decision maker (DM)

Table 6.2. Individual minimum and maximum for $\alpha = 0$ and 1

	$\alpha = 0$		$\alpha = 1$	
	minimum	maximum	minimum	maximum
$z_1(\boldsymbol{x})$	−21466.1025	18912.1711	−19523.6828	15819.0476
$z_2(\boldsymbol{x})$	−13860.5984	16745.1351	−12819.6667	13745.7143
$z_3(\boldsymbol{x})$	−16346.6667	11889.6949	−13203.3333	9767.5000

selects the initial value of the degree α to be 0.8 and the initial reference levels $(\bar{z}_1, \bar{z}_2, \bar{z}_3)$ to be $(-19000.0, -12500.0, -13000.0)$. Then the corresponding α-Pareto optimal solution together with the trade-off information can be obtained by solving the linear programming problem with the block angular structure using the Dantzig-Wolfe decomposition method, as shown in the second column of Table 6.3.

On the basis of such information, suppose that the DM updates the reference levels to $(-12000.0, -5700.0, -5900.0)$ improving the satisfaction levels for z_2 and z_3 at the expense of z_1. For the updated reference level, the corresponding results are shown in the third column of Table 6.3.

Table 6.3. Interactive processes for example

Interaction	1	2	3	4
α	0.8	0.8	0.7	0.7
$\bar{z}_1$	−19000.0	−12000.0	−12000.0	−11500.0
$\bar{z}_2$	−12500.0	−5700.0	−5800.0	−6100.0
$\bar{z}_3$	−13000.0	−5900.0	−6200.0	−6500.0
$z_1(\boldsymbol{x})$	−12029.1846	−11991.2295	−11991.2833	−11477.1369
$z_2(\boldsymbol{x})$	−5529.1846	−5691.2295	−5791.2833	−6077.1369
$z_3(\boldsymbol{x})$	−6029.1846	−5891.2295	−6191.2833	−6477.1369
$-\partial z_2(\boldsymbol{x})/\partial z_1(\boldsymbol{x})$	1.028227	1.028227	1.049640	1.049640
$-\partial z_3(\boldsymbol{x})/\partial z_1(\boldsymbol{x})$	1.153071	1.153071	1.182134	1.182134

In this way, at the fourth iteration, the satisficing solution for the DM is obtained as shown in the fifth column of Table 6.3 which becomes

$$x_1 = 0.0,\ x_2 = 0.0,\ x_3 = 51.1281,\ x_4 = 505.6407,\ x_5 = 647.7437,$$

$$x_6 = 0.0,\ x_7 = 1010.5668,\ x_8 = 7.0285,\ x_9 = 0.0,\ x_{10} = 0.0,$$

$$x_{11} = 0.0,\ x_{12} = 274.8343,\ x_{13} = 0.0,\ x_{14} = 0.0,\ x_{15} = 758.2781,$$

$$x_{16} = 0.0,\ x_{17} = 0.0,\ x_{18} = 0.0,\ x_{19} = 465.6783,\ x_{20} = 68.6434,$$

$$z_1(\boldsymbol{x}) = -11477.1369,\ z_2(\boldsymbol{x}) = -6077.1369,\ z_3(\boldsymbol{x}) = -6477.1369.$$

6.3 Interactive fuzzy multiobjective linear programming with fuzzy numbers

6.3.1 Fuzzy goals and solution concepts

As can be seen from the definition of α-Pareto optimality, in general, α-Pareto optimal solutions to the α-multiobjective linear programming problem (6.6) consist of an infinite number of points. The DM may be able to select a satisficing solution from an α-Pareto optimal solution set based on a subjective value judgment, making use of the interactive multiobjective linear programming with fuzzy numbers presented in the previous section.

However, considering the imprecise nature of the DM's judgments, it is natural to assume that the DM may have a fuzzy goal for each of the objective functions in the α-multiobjective linear programming problem (6.6). In a minimization problem, a goal stated by the DM may be to achieve "substantially less than or equal to some value p_i." This type of statement can be quantified by eliciting a corresponding membership function.

To elicit a membership function $\mu_i(\boldsymbol{c}_i\boldsymbol{x})$ from the DM for each of the objective functions $\boldsymbol{c}_i\boldsymbol{x}$, $i = 1, \ldots, k$, in the α-multiobjective linear programming problem, the individual minimum and maximum of each objective function are first calculated under the given constraints for $\alpha = 0$ and $\alpha = 1$. By taking account of the calculated individual minimum and maximum of each objective function for $\alpha = 0$ and $\alpha = 1$ together with the rate of increase of membership satisfaction, the DM may be able to determine the membership function $\mu_i(\boldsymbol{c}_i\boldsymbol{x})$ in a subjective manner which is a strictly monotone decreasing function with respect to $\boldsymbol{c}_i\boldsymbol{x}$.

So far we have restricted ourselves to a minimization problem and consequently assumed that the DM has a fuzzy goal such as "$\boldsymbol{c}_i\boldsymbol{x}$ should be substantially less than or equal to p_i." In the fuzzy approaches, we can further treat a more general case where the DM has two types of fuzzy goals, namely, fuzzy goals expressed in words such as "$\boldsymbol{c}_i\boldsymbol{x}$ should be in the vicinity of r_i" (called fuzzy equal) as well as "$\boldsymbol{c}_i\boldsymbol{x}$ should be substantially less than or equal to p_i" or "$\boldsymbol{c}_i\boldsymbol{x}$ should be substantially greater than or equal to q_i" (called fuzzy min or fuzzy max).

Such a generalized α-multiobjective linear programming problem may now be expressed as

$$\left.\begin{array}{rl} \text{fuzzy min} & z_i(\boldsymbol{x}) \quad i \in I_1 \\ \text{fuzzy max} & z_i(\boldsymbol{x}) \quad i \in I_2 \\ \text{fuzzy equal} & z_i(\boldsymbol{x}) \quad i \in I_3 \\ \text{subject to} & \boldsymbol{x} \in X(A, B, \boldsymbol{b}) \\ & (A, B, \boldsymbol{b}, \boldsymbol{c}) \in (\tilde{A}, \tilde{B}, \tilde{\boldsymbol{b}}, \tilde{\boldsymbol{c}})_\alpha \end{array}\right\} \tag{6.15}$$

where $I_1 \cup I_2 \cup I_3 = \{1, 2, \ldots, k\}$ and $I_i \cap I_j = \phi$, $i = 1,2,3$, $j = 1,2,3$, $i \neq j$.

To elicit a membership function $\mu_i(\boldsymbol{c}_i\boldsymbol{x})$ from the DM for a fuzzy goal like "$\boldsymbol{c}_i\boldsymbol{x}$ should be in the vicinity of r_i," it should be quite apparent that different functions can be utilized for both the left and right sides of r_i.

Concerning the membership functions of the problem (6.15), it is reasonable to assume that $\mu_i(\boldsymbol{c}_i\boldsymbol{x})$, $i \in I_1$, and the right side functions of $\mu_i(\boldsymbol{c}_i\boldsymbol{x})$, $i \in I_3$, are strictly monotone decreasing and continuous functions with respect to $\boldsymbol{c}_i\boldsymbol{x}$, while $\mu_i(\boldsymbol{c}_i\boldsymbol{x})$, $i \in I_2$, and the left side functions of $\mu_i(\boldsymbol{c}_i\boldsymbol{x})$, $i \in I_3$, are strictly monotone increasing and continuous functions with respect to $\boldsymbol{c}_i\boldsymbol{x}$.

When a fuzzy equal is included in the fuzzy goals of the DM, it is desirable that $\boldsymbol{c}_i\boldsymbol{x}$ should be as close to r_i as possible. Consequently, the notion of α-Pareto optimal solutions defined in terms of objective functions cannot be applied. For this reason, we introduce the concept of M-α-Pareto optimal solutions which is defined in terms of membership functions instead of objective functions, where M refers to membership.

Definition 6.3.1 (M-α-Pareto optimal solution).
$\boldsymbol{x}^* \in X(A^*, B^*, \boldsymbol{b}^*)$ for $(A^*, B^*, \boldsymbol{b}^*, \boldsymbol{c}^*) \in (\tilde{A}, \tilde{B}, \tilde{\boldsymbol{b}}, \tilde{\boldsymbol{c}})_\alpha$ is said to be an M-α-Pareto optimal solution to (6.15), *if and only if, there does not exist another $\boldsymbol{x} \in X(A, B, \boldsymbol{b})$, $(A, B, \boldsymbol{b}, \boldsymbol{c}) \in (\tilde{A}, \tilde{B}, \tilde{\boldsymbol{b}}, \tilde{\boldsymbol{c}})_\alpha$ such that $\mu_i(\boldsymbol{c}_i\boldsymbol{x}) \geq \mu_i(\boldsymbol{c}_i^*\boldsymbol{x}^*)$, $i = 1, \ldots, k$, with strict inequality holding for at least one i, where the corresponding values of parameters $(A^*, B^*, \boldsymbol{b}^*, \boldsymbol{c}^*) \in (\tilde{A}, \tilde{B}, \tilde{\boldsymbol{b}}, \tilde{\boldsymbol{c}})_\alpha$ are called M-α-level optimal parameters.*

Observe that the concept of M-α-Pareto optimal solutions defined in terms of membership functions is a natural extension to that of α-Pareto optimal solutions defined in terms of objective functions when a fuzzy equal is included in the fuzzy goals of the DM.

6.3.2 Minimax problems

Having elicited the membership functions from the DM for each of the objective functions, if a general aggregation function $\mu_D(\cdot)$ is introduced, a general multiobjective decision making problem can be defined by

$$\left.\begin{array}{ll} \text{maximize} & \mu_D(\mu_1(\boldsymbol{c}_1\boldsymbol{x}), \mu_2(\boldsymbol{c}_2\boldsymbol{x}), \ldots, \mu_k(\boldsymbol{c}_k\boldsymbol{x}), \alpha) \\ \text{subject to} & \boldsymbol{x} \in X(A, B, \boldsymbol{b}) \\ & (A, B, \boldsymbol{b}, \boldsymbol{c}) \in (\tilde{A}, \tilde{B}, \tilde{\boldsymbol{b}}, \tilde{\boldsymbol{c}})_\alpha. \end{array}\right\} \quad (6.16)$$

where $\mu_D(\cdot)$ is assumed to be strictly increasing and continuous with respect to $\mu_i(\cdot)$ and α.

Probably the most crucial problem in this case is the identification of an appropriate aggregation function which well represents the human DM's fuzzy preferences. If $\mu_D(\cdot)$ can be explicitly identified, then the problem is reduced to a standard mathematical programming problem. However, this rarely happens, and as an alternative approach, it becomes evident that interaction with the DM is necessary.

To generate a candidate for the satisficing solution, which is also M-α-Pareto optimal, in our decision making method, the DM is asked to specify the degree α of the α-level set and the reference membership levels. Observe that the idea of the reference membership levels can be viewed as an obvious extension of the idea of the reference point in Wierzbicki [135, 136].

Once the DM's degree α and reference membership levels $\bar{\mu}_i$, $i = 1, \ldots, k$, are specified, the corresponding M-α-Pareto optimal solution, which is, in the minimax sense, nearest to the requirement or better than that if the reference levels are attainable, is obtained by solving the following minimax problem:

$$\left.\begin{array}{ll} \text{minimize} & \displaystyle\max_{i=1,\ldots,k} (\bar{\mu}_i - \mu_i(\boldsymbol{c}_i\boldsymbol{x})) \\ \text{subject to} & \boldsymbol{x} \in X(A, B, \boldsymbol{b}) \\ & (A, B, \boldsymbol{b}, \boldsymbol{c}) \in (\tilde{A}, \tilde{B}, \tilde{\boldsymbol{b}}, \tilde{\boldsymbol{c}})_\alpha \end{array}\right\} \quad (6.17)$$

or equivalently

$$\left.\begin{array}{ll} \text{minimize} & v \\ \text{subject to} & \bar{\mu}_i - \mu_i(\boldsymbol{c}_i\boldsymbol{x}) \leq v, \;\; i = 1, \ldots, k \\ & \boldsymbol{x} \in X(A, B, \boldsymbol{b}) \\ & (A, B, \boldsymbol{b}, \boldsymbol{c}) \in (\tilde{A}, \tilde{B}, \tilde{\boldsymbol{b}}, \tilde{\boldsymbol{c}})_\alpha. \end{array}\right\} \quad (6.18)$$

However, with the strictly monotone decreasing or increasing membership function, which may be nonlinear, the resulting problem becomes a nonlinear programming problem. For notational convenience, denote the strictly monotone decreasing functions of $\mu_i(\boldsymbol{c}_i\boldsymbol{x})$, $i \in I_1$ and the right side functions of $\mu_i(\boldsymbol{c}_i\boldsymbol{x})$, $i \in I_3$ by $d_{iR}(\boldsymbol{c}_i\boldsymbol{x})$, $i \in I_1 \cup I_3$, and strictly monotone increasing functions of $\mu_i(\boldsymbol{c}_i\boldsymbol{x})$, $i \in I_2$ and the left side functions of $\mu_i(\boldsymbol{c}_i\boldsymbol{x})$, $i \in I_3$ by $d_{iL}(\boldsymbol{c}_i\boldsymbol{x})$ ($i \in I_2 \cup I_3$). Then in order to solve the formulated problem on the basis of the linear programming and Dantzig-Wolfe decomposition method, using the strictly monotone property of $d_{iL}(\cdot)$ and $d_{iR}(\cdot)$, we first convert each constraint $\bar{\mu}_i - \mu_i(\boldsymbol{c}_i\boldsymbol{x}) \leq v$, $i = 1, \ldots, k$, of the minimax problem (6.18) as follows:

$$\left.\begin{array}{rl} \text{minimize} & v \\ \text{subject to} & \boldsymbol{c}_i\boldsymbol{x} \leq d_{iR}^{-1}(\bar{\mu}_i - v), \;\; i \in I_1 \cup I_3 \\ & \boldsymbol{c}_i\boldsymbol{x} \geq d_{iL}^{-1}(\bar{\mu}_i - v), \;\; i \in I_2 \cup I_3 \\ & \boldsymbol{x} \in X(A, B, \boldsymbol{b}) \\ & (A, B, \boldsymbol{b}, \boldsymbol{c}) \in (\tilde{A}, \tilde{B}, \tilde{\boldsymbol{b}}, \tilde{\boldsymbol{c}})_\alpha \end{array}\right\} \qquad (6.19)$$

where d_{iR}^{-1} and d_{iL}^{-1} are pseudo-inverse defined by

$$d_{iR}^{-1}(h) = \sup\{y \mid d_{iR}(y) \geq h\}, \qquad (6.20)$$

$$d_{iL}^{-1}(h) = \inf\{y \mid d_{iL}(y) \geq h\}. \qquad (6.21)$$

In this formulation, however, constraints are nonlinear because the parameters A, B, $\boldsymbol{b}$, $\boldsymbol{c}$ and $\boldsymbol{d}$ are treated as decision variables. To deal with such nonlinearities, we introduce the following set-valued functions $S_{iR}(\cdot)$, $S_{iL}(\cdot)$, $T_j(\cdot,\cdot)$ and $U(\cdot,\cdot)$.

$$\left.\begin{array}{rcl} S_{iR}(\boldsymbol{c}_i) &=& \{(\boldsymbol{x}, v) \mid \boldsymbol{c}_i\boldsymbol{x} \leq d_{iR}^{-1}(\bar{\mu}_i - v), \;\; i \in I_1 \cup I_3, \;\; \boldsymbol{x} \geq \boldsymbol{0}\} \\ S_{iL}(\boldsymbol{c}_i) &=& \{(\boldsymbol{x}, v) \mid \boldsymbol{c}_i\boldsymbol{x} \geq d_{iL}^{-1}(\bar{\mu}_i - v), \;\; i \in I_2 \cup I_3, \;\; \boldsymbol{x} \geq \boldsymbol{0}\} \\ T_j(B_j, \boldsymbol{b}_j) &=& \{\boldsymbol{x}_j \mid B_j\boldsymbol{x}_j \leq \boldsymbol{b}_j, \;\; \boldsymbol{x}_j \geq \boldsymbol{0}\} \\ U(A, \boldsymbol{b}_0) &=& \{\boldsymbol{x} \mid A\boldsymbol{x} \leq \boldsymbol{b}_0, \;\; \boldsymbol{x} \geq \boldsymbol{0}\} \end{array}\right\} \qquad (6.22)$$

It can be verified that the following relations hold for $S_{iR}(\cdot)$, $S_{iL}(\cdot)$, $T_j(\cdot,\cdot)$ and $U(\cdot,\cdot)$ when $\boldsymbol{x} \geq \boldsymbol{0}$.

Proposition 6.3.1.

(1) $\boldsymbol{c}_i^1 \leq \boldsymbol{c}_i^2 \Rightarrow S_{iR}(\boldsymbol{c}_i^1) \supseteq S_{iR}(\boldsymbol{c}_i^2)$, $S_{iL}(\boldsymbol{c}_i^1) \subseteq S_{iL}(\boldsymbol{c}_i^2)$

(2) $B_j^1 \leq B_j^2 \Rightarrow T_j(B_j^1, \boldsymbol{b}_j) \supseteq T_j(B_j^2, \boldsymbol{b}_j)$

(3) $\boldsymbol{b}_j^1 \leq \boldsymbol{b}_j^2 \Rightarrow T_j(B_j, \boldsymbol{b}_j^1) \subseteq T_j(B_j, \boldsymbol{b}_j^2)$

(4) $A^1 \leq A^2 \Rightarrow U(A^1, \boldsymbol{b}_0) \supseteq U(A^2, \boldsymbol{b}_0)$

(5) $\boldsymbol{b}_0^1 \leq \boldsymbol{b}_0^2 \Rightarrow U(A, \boldsymbol{b}_0^1) \subseteq U(A, \boldsymbol{b}_0^2)$

Using the properties of the α-level sets for the matrices of the fuzzy numbers A_i, B_i and the vectors of the fuzzy numbers $\boldsymbol{b}_{h'}$, $\boldsymbol{c}_j$, the feasible regions for A_i, B_i, $\boldsymbol{b}_{h'}$ and $\boldsymbol{c}_j$ can be denoted respectively by the closed intervals $[A_{i\alpha}^L, A_{i\alpha}^R]$, $[B_{i\alpha}^L, B_{i\alpha}^R]$, $[\boldsymbol{b}_{h'\alpha}^L, \boldsymbol{b}_{h'\alpha}^R]$, $[\boldsymbol{c}_{j\alpha}^L, \boldsymbol{c}_{j\alpha}^R]$, respectively.

Consequently, using the results in Proposition 6.3.1, an optimal solution can be obtained by solving the following problem:

$$\left.\begin{array}{ll}
\text{minimize} & v \\
\text{subject to} & \boldsymbol{c}_{i1\alpha}^L \boldsymbol{x}_1 + \cdots + \boldsymbol{c}_{ip\alpha}^L \boldsymbol{x}_p \leq d_{iR}^{-1}(\bar{\mu}_i - v),\ i \in I_1 \cup I_3 \\
& \boldsymbol{c}_{i1\alpha}^R \boldsymbol{x}_1 + \cdots + \boldsymbol{c}_{ip\alpha}^R \boldsymbol{x}_p \geq d_{iL}^{-1}(\bar{\mu}_i - v),\ i \in I_2 \cup I_3 \\
& A_{1\alpha}^L \boldsymbol{x}_1 + \cdots + A_{p\alpha}^L \boldsymbol{x}_p \leq \boldsymbol{b}_{0\alpha}^R \\
& B_{1\alpha}^L \boldsymbol{x}_1 \qquad\qquad\qquad \leq \boldsymbol{b}_{1\alpha}^R \\
& \qquad \ddots \qquad\qquad\qquad\quad \vdots \\
& \qquad\qquad\qquad B_{p\alpha}^L \boldsymbol{x}_p \leq \boldsymbol{b}_{p\alpha}^R \\
& \boldsymbol{x}_j \geq \boldsymbol{0},\ j = 1, \ldots, p.
\end{array}\right\} \tag{6.23}$$

It is important to note here that, if the value of v is fixed, this formulation can be reduced to a set of linear inequalities. Obtaining the optimal solution v^* to the above problem is equivalent to determining the minimum value of v so that there exists an admissible set satisfying the constraints of (6.23). Since v satisfies $\bar{\mu}_{\max} - 1 \leq v \leq \bar{\mu}_{\max}$, where $\bar{\mu}_{\max}$ denotes the maximum value of $\bar{\mu}_i$, $i = 1, \ldots, k$, the following method can be used for solving this problem by combined use of the bisection method and the simplex method of linear programming.

(1) Set $v = \bar{\mu}_{\max}$ and test whether an admissible set satisfying the constraints of (6.23) exists or not by making use of phase one of the simplex method. If an admissible set exists, proceed. Otherwise, the DM must reassess the membership function.

(2) Set $v = \bar{\mu}_{\max} - 1$ and test whether an admissible set satisfying the constraints of (6.23) exists or not using phase one of the simplex method. If an admissible set exists, set $v^* = \bar{\mu}_{\max} - 1$. Otherwise, go to the next step since the minimum v which satisfies the constraints of (6.23) exists between $\bar{\mu}_{\max} - 1$ and $\bar{\mu}_{\max}$.

(3) For the initial value of $v = \bar{\mu}_{\max} - 0.5$, update the value of v using the bisection method as follows:

$$\begin{cases} v_{n+1} = v_n - 1/2^{n+1} & \text{if an admissible set exists for } v_n, \\ v_{n+1} = v_n + 1/2^{n+1} & \text{if no admissible set exists for } v_n. \end{cases}$$

For each v_n, $n = 1, 2, \ldots$, test whether an admissible set of (6.23) exists or not using the sensitivity analysis technique for the changes in the right-hand side of the simplex method and determine the minimum value of v satisfying the constraints of (6.23).

In this way, we can determine the optimal solution v^*. Then the DM selects an appropriate standing objective from among the objectives $\boldsymbol{c}_i\boldsymbol{x}$, $i = 1, \ldots, k$. For notational convenience in the following without loss of generality, let it be $\boldsymbol{c}_{1\alpha}^L \boldsymbol{x}$ and $1 \in I_1$. Then the following linear programming problem is solved for $v = v^*$:

$$\left.\begin{array}{ll}\text{minimize} & \boldsymbol{c}_{1\alpha}^L \boldsymbol{x} \\ \text{subject to} & \boldsymbol{c}_{i1\alpha}^L \boldsymbol{x}_1 + \cdots + \boldsymbol{c}_{ip\alpha}^L \boldsymbol{x}_p \leq d_{iR}^{-1}(\bar{\mu}_i - v^*),\ i \in I_1 \cup I_3, i \neq 1 \\ & \boldsymbol{c}_{i1\alpha}^R \boldsymbol{x}_1 + \cdots + \boldsymbol{c}_{ip\alpha}^R \boldsymbol{x}_p \leq d_{iL}^{-1}(\bar{\mu}_i - v^*),\ i \in I_2 \cup I_3, i \neq 1 \\ & A_{1\alpha}^L \boldsymbol{x}_1 + \cdots + A_{p\alpha}^L \boldsymbol{x}_p \leq \boldsymbol{b}_{0\alpha}^R \\ & B_{1\alpha}^L \boldsymbol{x}_1 \qquad\qquad\qquad \leq \boldsymbol{b}_{1\alpha}^R \\ & \qquad\qquad \ddots \qquad\qquad\quad \vdots \\ & \qquad\qquad\qquad B_{p\alpha}^L \boldsymbol{x}_p \leq \boldsymbol{b}_{p\alpha}^R \\ & \boldsymbol{x}_j \geq \boldsymbol{0},\ j = 1, \ldots, p. \end{array}\right\} \tag{6.24}$$

For convenience in our subsequent discussion, we assume that the optimal solution $\boldsymbol{x}^*$ to (6.24) satisfies the following conditions:

$$\boldsymbol{c}_{i\alpha}^L \boldsymbol{x}^* \triangleq \boldsymbol{c}_{i1\alpha}^L \boldsymbol{x}_1^* + \cdots + \boldsymbol{c}_{ip\alpha}^L \boldsymbol{x}_p^* = d_{iR}^{-1}(\bar{\mu}_i - v^*),\ i \in I_1 \cup I_{3R},$$

$$\boldsymbol{c}_{i\alpha}^R \boldsymbol{x}^* \triangleq \boldsymbol{c}_{i1\alpha}^R \boldsymbol{x}_1^* + \cdots + \boldsymbol{c}_{ip\alpha}^R \boldsymbol{x}_p^* = d_{iL}^{-1}(\bar{\mu}_i - v^*),\ i \in I_2 \cup I_{3L}.$$

where $I_3 = I_{3L} \cup I_{3R}$ and $I_{3L} \cap I_{3R} = \emptyset$.

The relationships between the optimal solutions to (6.23) and the M-α-Pareto optimal concept of the problem (6.15) can be characterized by the following theorems [78, 91]:

Theorem 6.3.1. *If $\boldsymbol{x}^*$ is a unique optimal solution to* (6.23), *then $\boldsymbol{x}^*$ is an M-α-Pareto optimal solution to* (6.15).

Theorem 6.3.2. *If $\boldsymbol{x}^*$ is an M-α-Pareto optimal solution to* (6.15) *and* $(A^*, B^*, \boldsymbol{b}^*, \boldsymbol{c}^*)$ *are α-level optimal parameters, then $\boldsymbol{x}^*$ is an optimal solution to* (6.23) *for some* $\bar{\boldsymbol{\mu}} = (\bar{\mu}_1, \ldots, \bar{\mu}_k)$.

It must be observed here that for generating M-α-Pareto optimal solutions using Theorem 6.3.1, the uniqueness of the solution must be verified. In the ad hoc numeral approach, however, to test the M-α-Pareto optimality of a current optimal solution $\boldsymbol{x}^*$, we formulate and solve the following linear programming problem:

$$\left.\begin{array}{ll} \text{maximize} & \varepsilon = \sum_{i=1}^{k} \varepsilon_i \\ \text{subject to} & \boldsymbol{c}_{i\alpha}^L \boldsymbol{x} + \varepsilon_i = \boldsymbol{c}_{i\alpha}^L \boldsymbol{x}^*, \; i \in I_1 \cup I_3 \\ & \boldsymbol{c}_{i\alpha}^R \boldsymbol{x} - \varepsilon_i = \boldsymbol{c}_{i\alpha}^R \boldsymbol{x}^*, \; i \in I_2 \cup I_3 \\ & A_\alpha^L \boldsymbol{x} \leq \boldsymbol{b}_{0\alpha}^R \\ & B_{j\alpha}^L \boldsymbol{x}_j \leq \boldsymbol{b}_{j\alpha}^R, \; j = 1, \ldots, p \\ & \boldsymbol{x}_j \geq \boldsymbol{0}, \; j = 1, \ldots, p \\ & \varepsilon_i \geq 0, \; i = 1, \ldots, k. \end{array}\right\} \quad (6.25)$$

For the optimal solution $\bar{\boldsymbol{x}}$, $\bar{\boldsymbol{\varepsilon}}$ to the problem (6.25) involving $\boldsymbol{x}$ and ε_i as decision variables, the following theorems hold:

Theorem 6.3.3. *For the optimal solution $\bar{\boldsymbol{x}}$, $\bar{\boldsymbol{\varepsilon}}$ to the problem* (6.25),

(1) *If $\bar{\boldsymbol{\varepsilon}} = 0$, then $\boldsymbol{x}^*$ is an M-α-Pareto optimal solution to* (6.15).
(2) *If $\bar{\boldsymbol{\varepsilon}} \neq 0$, then $\boldsymbol{x}^*$ is not an M-α-Pareto optimal solution, but $\bar{\boldsymbol{x}}$ is an M-α-Pareto optimal solution to* (6.15).

It should be noted here that the M-α-Pareto optimality test problem (6.25) also preserves both the linearity and block angular structure, and hence the Dantzig-Wolfe decomposition can be used for solving it.

6.3.3 Interactive fuzzy programming

Now, given the M-α-Pareto optimal solution for the degree α and the reference membership levels specified by the DM by solving the corresponding minimax problem, the DM must either be satisfied with the current M-α-Pareto optimal solution or act on this solution by updating the reference membership levels and/or the degree α. To help the DM express a degree of preference, the trade-off information between a standing membership function and each of the other membership functions as well as between the degree α and the membership functions is very useful. Such trade-off information is easily obtainable since it is closely related to the simplex multipliers of the problem (6.23).

To derive the trade-off information, define the following Lagrangian function L for the problem (6.23) with respect to the coupling constraints:

$$\begin{aligned} L = c_{1\alpha}^L x + \sum_{i \in I_1 \cup I_{3R}} \pi_{iR}\{c_{i\alpha}^L x - d_{iR}^{-1}(\bar{\mu}_i - v^*)\} \\ + \sum_{i \in I_2 \cup I_{3L}} \pi_{iL}\{d_{iL}^{-1}(\bar{\mu}_i - v^*) - c_{i\alpha}^R x\} + \lambda(A_\alpha^L x - b_{0\alpha}^R). \end{aligned} \tag{6.26}$$

where π_{iR}, π_{iL} and λ are simplex multipliers corresponding to the constraints of the problem (6.23).

Here it is assumed that the problem (6.23) has a unique and nondegenerate optimal solution satisfying the following conditions:

(1) $\pi_{iR} > 0,\ i \in I_1 \cup I_{3R};\ i \neq 1$
(2) $\pi_{iL} > 0,\ i \in I_2 \cup I_{3L}$.

Then by using the results in Haimes and Chankong [31], the following expression holds:

$$-\frac{\partial(c_{1\alpha}^L x)}{\partial(c_{i\alpha}^L x)} = \pi_{iR}, \quad i \in I_1 \cup I_{3R} \tag{6.27}$$

$$-\frac{\partial(c_{1\alpha}^L x)}{\partial(c_{i\alpha}^R x)} = -\pi_{iL}, \quad i \in I_2 \cup I_{3L}. \tag{6.28}$$

Furthermore, using the strictly monotone decreasing or increasing property of $d_{iR}(\cdot)$ or $d_{iL}(\cdot)$ together with the chain rule, if $d_{iR}(\cdot)$ and $d_{iL}(\cdot)$ are differentiable at the optimal solution to (6.23), it holds that:

$$-\frac{\partial \mu_1(c_{1\alpha}^L x)}{\partial \mu_i(c_{i\alpha}^L x)} = \frac{d'_{1R}(c_{1\alpha}^L x)}{d'_{iR}(c_{i\alpha}^L x)} \pi_{iR}, \quad i \in I_1 \cup I_{3R};\ i \neq 1 \tag{6.29}$$

$$-\frac{\partial \mu_1(c_{1\alpha}^L x)}{\partial \mu_i(c_{i\alpha}^R x)} = -\frac{d'_{1R}(c_{1\alpha}^L x)}{d'_{iL}(c_{i\alpha}^R x)} \pi_{iL}, \quad i \in I_2 \cup I_{3L}. \tag{6.30}$$

where $d'_{iR}(\cdot)$ and $d'_{iL}(\cdot)$ denote the differential coefficients of $d_{iR}(\cdot)$ and $d_{iL}(\cdot)$, respectively.

Regarding a trade-off rate between $\mu_1(c_{1\alpha}^L x)$ and α, the following relation holds based on the sensitivity theorem [21]:

$$\begin{aligned} \frac{\partial \mu_1(c_{1\alpha}^L x)}{\partial \alpha} = d'_{1R}(c_{1\alpha}^L x) \left\{ \frac{\partial(c_{1\alpha}^L)}{\partial \alpha} x + \sum_{i \in I_1 \cup I_{3R}} \pi_{iR} \frac{\partial(c_{i\alpha}^L)}{\partial \alpha} x \right. \\ \left. - \sum_{i \in I_2 \cup I_{3L}} \pi_{iL} \frac{\partial(c_{i\alpha}^R)}{\partial \alpha} x + \lambda \left\{ \frac{\partial(A_{j\alpha}^L)}{\partial \alpha} x - \frac{\partial(b_{j\alpha}^R)}{\partial \alpha} \right\} \right\}. \end{aligned} \tag{6.31}$$

It should be noted that to obtain the trade-off rate information from (6.27) to (6.31), all the constraints of problem (6.23) must be active for the current optimal solution. Therefore, if there are inactive constraints, it is necessary to replace $\bar{\mu}_j$ for inactive constraints by $d_{iR}(\boldsymbol{c}_{i\alpha}^L \boldsymbol{x}^*) + v^*$ or $d_{iL}(\boldsymbol{c}_{i\alpha}^R \boldsymbol{x}^*) + v^*$ and solve the corresponding problem (6.23) for obtaining the simplex multipliers.

Now, following the preceding discussions, we can present the interactive algorithm to derive the satisficing solution for the DM from the M-α-Pareto optimal solution set. The steps marked with an asterisk involve interaction with the DM.

Interactive fuzzy multiobjective linear programming with fuzzy numbers

Step 0: By applying the Dantzig-Wolfe decomposition method, calculate the individual minimum and maximum of each objective function under the given constraints for $\alpha = 0$ and $\alpha = 1$.

Step 1*: Elicit a membership function from the DM for each of the objective functions.

Step 2*: Ask the DM to select the initial value of α $(0 \leq \alpha \leq 1)$ and set the initial reference membership levels $\bar{\mu}_i = 1$, $i = 1, \ldots, k$.

Step 3: For the degree α and the reference membership levels specified by the DM, using the Dantzig-Wolfe decomposition method, solve the minimax problem and perform the M-α-Pareto optimality test to obtain the M-α-Pareto optimal solution and the trade-off rates between the membership functions and the degree α.

Step 4*: The DM is supplied with the corresponding M-α-Pareto optimal solution and the trade-off rates between the membership functions and the degree α. If the DM is satisfied with the current membership function values of the M-α-Pareto optimal solution and α, stop. Otherwise, the DM must update the reference membership levels and/or the degree α by considering the current values of the membership functions and α together with the trade-off rates between the membership functions and the degree α and return to Step 3.

It should be stressed to the DM that (1) any improvement of one membership function can be achieved only at the expense of at least one of the other membership functions for some fixed degree α and (2) the greater value of the degree α gives the worse values of the membership functions for some fixed reference membership levels.

6.3.4 Numerical example

To illustrate the proposed method, consider the following three-objective linear programming problem with fuzzy numbers:

$$\begin{aligned}
\text{fuzzy max } z_1 &= \tilde{c}_{11}x_1 + \tilde{c}_{12}x_2 + \tilde{c}_{13}x_3 + 3.8x_4 + 4.2x_5 + 4.5x_6 \\
&\quad + \tilde{c}_{17}x_7 + \tilde{c}_{18}x_8 + \tilde{c}_{19}x_9 + 3.2x_{10} + 3.5x_{11} + 4.2x_{12} \\
\text{fuzzy equal } z_2 &= 29.5x_1 + 28x_2 + 26.3x_3 + \tilde{c}_{24}x_4 + \tilde{c}_{25}x_5 + \tilde{c}_{26}x_6 \\
&\quad + 28.4x_7 + 27x_8 + 25.3x_9 + \tilde{c}_{210}x_{10} + \tilde{c}_{211}x_{11} + \tilde{c}_{212}x_{12} \\
\text{fuzzy min } z_3 &= \tilde{c}_{31}x_1 + 25x_2 + 30x_3 + \tilde{c}_{34}x_4 + 19x_5 + 9x_6 \\
&\quad + 8x_7 + 18x_8 + 28x_9 + \tilde{c}_{310}x_{10} + \tilde{c}_{311}x_{11} + 5x_{12}
\end{aligned}$$

$$\begin{aligned}
\text{subject to } & 12x_1 + 13x_2 + 17x_3 + 19x_4 + 23x_5 + 29x_6 \\
& +31x_7 + 37x_8 + 41x_9 + 43x_{10} + 47x_{11} + 51x_{12} \leq 2525.2 \\
& x_1 + 2x_2 + 3x_3 + 5x_4 + 7x_5 + 11x_6 \\
& +13x_7 + 17x_8 + 19x_9 + 23x_{10} + 29x_{11} + 31x_{12} \leq 1158.7 \\
& 12x_1 + 2x_2 + x_3 \leq 98.5 \\
& 2x_1 + 21x_2 + 3x_3 \leq 192.3 \\
& 30x_1 + 20x_2 + 300x_3 \leq 2874 \\
& 49x_1 + 55x_2 + 47x_3 \leq 1021 \\
& 49x_4 + 6x_5 + 7x_6 \leq 525 \\
& 19x_4 + 210x_5 + 18x_6 \leq 2120 \\
& x_4 + x_5 + 10x_6 \leq 98.7 \\
& 74x_4 + 76x_5 + 78x_6 \leq 1485 \\
& 103x_7 + 12x_8 + 9x_9 \leq 1026.9 \\
& 20x_7 + 246x_8 + 20x_9 \leq 2052.8 \\
& 2x_7 + 3x_8 + 32x_9 \leq 297.8 \\
& 49x_7 + 48x_8 + 47x_9 \leq 1025.4 \\
& 23x_{10} + 2x_{11} + x_{12} \leq 249.7 \\
& 98x_{10} + 2340x_{11} + 120x_{12} \leq 23950 \\
& 4x_{10} + 5x_{11} + 100x_{12} \leq 987.6 \\
& 123x_{10} + 124x_{11} + 125x_{12} \leq 2500 \\
& x_i \geq 0 \quad i = 1, \ldots, 12.
\end{aligned}$$

The fuzzy numbers involved in this example are explained in Table 6.4.

Having calculated the individual minimum and maximum of each of the objective functions for $\alpha = 0$ and $\alpha = 1$, assume that the interaction with the DM establishes the following membership functions and the corresponding assessment values:

Table 6.4. Fuzzy numbers for example

$\tilde{m}$	a,	m,	b	Left	Right
$\tilde{c}_{11}$	(1.20 ,	3.20 ,	5.20)	exponential	linear
$\tilde{c}_{12}$	(1.70 ,	3.70 ,	5.70)	linear	linear
$\tilde{c}_{13}$	(2.20 ,	4.20 ,	6.20)	linear	linear
$\tilde{c}_{17}$	(2.00 ,	4.00 ,	6.00)	linear	exponential
$\tilde{c}_{18}$	(2.00 ,	4.00 ,	6.00)	linear	linear
$\tilde{c}_{19}$	(4.00 ,	6.00 ,	8.00)	linear	linear
$\tilde{c}_{24}$	(24.00,	27.00,	30.00)	exponential	exponential
$\tilde{c}_{25}$	(21.50,	24.50,	27.50)	linear	linear
$\tilde{c}_{26}$	(20.20,	23.20,	26.20)	linear	linear
$\tilde{c}_{210}$	(26.00,	29.00,	32.00)	linear	linear
$\tilde{c}_{211}$	(23.90,	26.90,	29.90)	linear	linear
$\tilde{c}_{212}$	(21.80,	24.80,	27.80)	exponential	exponential
$\tilde{c}_{31}$	(8.00 ,	10.00,	12.00)	linear	linear
$\tilde{c}_{34}$	(25.00,	29.00,	34.00)	exponential	exponential
$\tilde{c}_{310}$	(14.00,	17.00,	20.00)	linear	exponential
$\tilde{c}_{311}$	(10.00,	13.00,	16.00)	linear	linear

$$\begin{aligned}
\text{fuzzy max } z_1 &: \text{exponential}(z_1^0, z_1^{0.5}, z_1^1) = (200, 250, 330) \\
\text{fuzzy equal } z_2 &: \text{linear}(z_{2L}^0, z_{2L}^1 = z_{2R}^1, z_{2R}^0) = (500, 1000, 1500) \\
\text{fuzzy min } z_3 &: \text{exponential}(z_3^1, z_3^{0.5}, z_3^0) = (400, 800, 1000)
\end{aligned}$$

where, within the interval $[z_i^0, z_i^1]$, the linear and exponential membership functions are defined as [78]:

$$\mu_i(z_i(\boldsymbol{x})) = \frac{z_i(\boldsymbol{x}) - z_i^0}{z_i^1 - z_i^0}$$

$$\mu_i(z_i(\boldsymbol{x})) = a_i \left[1 - \exp\left\{-\alpha_i \frac{z_i(\boldsymbol{x}) - z_i^0}{z_i^1 - z_i^0}\right\}\right],$$

where $a_i > 1, \alpha_i > 0$ or $a_i < 0, \alpha_i < 0$.

The minimax problem is solved for the degree $\alpha = 0.9$ specified by the DM and the initial membership values 1, and the DM is supplied with the corresponding M-α-Pareto optimal solution and the trade-off rates between the membership functions and the degree α as shown in the second column of Table 6.5.

Since the DM is not satisfied with the current membership values, the DM updates the reference membership levels as (0.5, 0.9, 0.7) for improving μ_2 at the expense of μ_1. The same procedures continue in this fashion, and in this example, at the fifth iteration, the satisficing solution for the DM is derived and all the interactive processes are summarized in Table 6.5.

Table 6.5. Interactive processes for example

Interaction	1	2	3	4	5
$\bar{\mu}_1$	1	0.5	0.6	0.6	0.5
$\bar{\mu}_2$	1	0.9	0.9	0.8	0.8
$\bar{\mu}_3$	1	0.7	0.7	0.5	0.6
α	0.9	0.8	0.7	0.6	0.6
μ_1	0.41	0.28	0.36	0.43	0.39
μ_2	0.41	0.68	0.66	0.63	0.69
μ_3	0.41	0.48	0.46	0.37	0.49
z_1	240.01	225.72	233.56	241.80	236.79
z_2	1292.86	1159.94	1172.27	1184.93	1157.37
z_3	841.56	810.09	822.27	867.96	807.50
x_1	0.00	0.00	0.00	0.00	0.00
x_2	0.00	0.00	0.00	0.00	0.00
x_3	7.70	7.57	9.58	9.58	9.58
x_4	0.00	0.00	0.00	0.00	0.00
x_5	8.33	9.33	6.22	9.33	5.23
x_6	8.94	8.94	9.25	8.94	9.35
x_7	6.90	2.44	3.75	2.07	4.27
x_8	0.00	0.00	0.00	0.00	0.00
x_9	8.87	9.15	9.07	9.18	9.04
x_{10}	0.00	0.00	0.00	0.00	0.00
x_{11}	0.00	0.00	0.00	0.00	0.00
x_{12}	9.88	9.88	9.87	9.87	9.88
$-\partial\mu_2/\partial\mu_1$	1.56	1.30	1.29	1.30	1.25
$-\partial\mu_3/\partial\mu_1$	11.87	11.89	18.80	81.17	67.63
$-\partial\mu_1/\partial\alpha$	0.63	0.54	0.55	0.49	0.53

6.3.5 Numerical experiments

Through a lot of computational experiments on numerical examples with both 50 and 200 variables, the advantages of the Dantzig-Wolfe decomposition method over the conventional method are discussed with respect to processing time and required memory storage.

Experimental environment. Our experiments were performed on a workstation, Sun SPARC Station 10 (main memory: 32MB) using a GNU C compiler (version 2.7). Problems with 3 objectives and 15 coupling constraints were solved as numerical examples, where coefficients in them were determined in accordance with the following rules.

(1) Each element of $\boldsymbol{c}_i$, A_i and B_i is given a certain value chosen from $(-5, 5)$ randomly.
(2) For $\boldsymbol{c}_i$, A_i and B_i generated in (1), each element of $\boldsymbol{b}_0$ and $\boldsymbol{b}_i$ is given a certain value chosen from $(5000, 10000)$ randomly so that the feasible region may exist.

On the basis of these values, assuming triangular fuzzy numbers for coefficients and linear membership functions for fuzzy goals, the resulting linear programming problems with block angular structures were solved.

The content of experiments. The following three methods to solve linear programming problems with block angular structures were compared with respect to the processing time and required memory capacity.

(1) **Decomposition-based method**
By the Dantzig-Wolfe decomposition method, a given problem is solved after dividing the original problem into one master problem and n independent subproblems.
(2) **Simplex-based method**
A given problem is solved by the conventional revised simplex method, when all coefficients that don't appear in block constraints are considered to be 0 as shown in Figure 6.1. As seen from Figure 6.1, in the direct

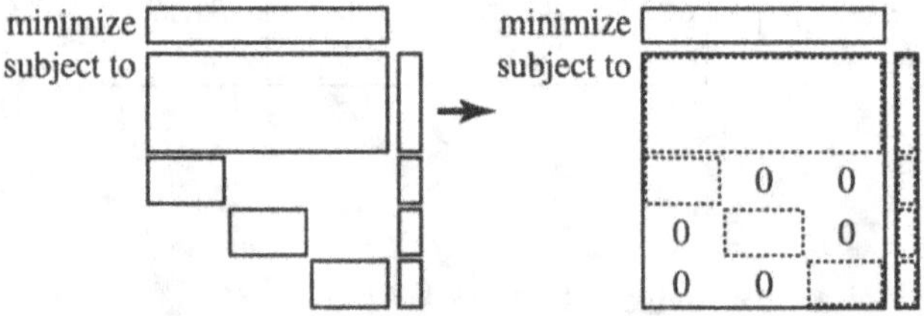

Figure 6.1. Problem conversion in (2).

application of the revised simplex method, extra memory capacity is needed for coefficients set to 0 that don't appear in the original problem.
(3) **Improved simplex-based method**
In order to reduce the redundant data processing in method (2), an im-

proved simplex method in consideration of the sparseness of the coefficient matrix is applied.

To be more specific, in the following procedure of the simplex method:

Step 1: Using relative cost coefficients $\bar{c}_j$, find s such that $\min_{\bar{c}_j<0} \bar{c}_j = \bar{c}_s$. If $\bar{c}_s \geq 0$, then terminate the procedure since the optimal solution is obtained.

Step 2: If $\bar{a}_{is} \leq 0$ for every i, then terminate the procedure since it shows that the minimal value is not unbounded.

Step 3: If there exists at least one positive $\bar{a}_{is}$, find r such that

$$\min_{\bar{a}_{is}>0} \frac{\bar{b}_i}{\bar{a}_{is}} = \frac{\bar{b}_r}{\bar{a}_{rs}} = \theta.$$

Step 4: Perform the pivot operation on $\bar{a}_{rs}$.

Step 2 and Step 3 are changed as follows.

Step 2: Now let the coupling constraints be from the m_0th row to the n_0th row and the block constraints concerning the sth column be from the m_sth row to the n_sth row. If $\bar{a}_{is} \leq 0$ for every i such that $m_0 < i < n_0$ or $m_s < i < n_s$, then terminate the procedure since it shows that the minimal value is not unbounded.

Step 3: If there exists at least one positive $\bar{a}_{is}$, find r such that

$$\min_{\bar{a}_{is}>0} \frac{\bar{b}_i}{\bar{a}_{is}} = \frac{\bar{b}_r}{\bar{a}_{rs}} = \theta.$$

Three methods mentioned above were compared concerning the processing time and required memory capacity. To be more specific:

(1) **Processing time**
- See relations between the number of blocks and the processing time in the application of them to problems involving 20 and 200 variables.
- Consider the case when variables are not shared on the average among blocks, especially, a certain block are allocated prominently many variables in a problem with 200 variables and 20 blocks. In this case, see relations between the number of variables in the prominent block and the processing time.

(2) **Required memory capacity**
- See relations between the number of blocks and the required memory capacity for problems with 2000 variables.

Results and discussions.

Processing time. Look into the influence of the number of blocks included in a problem on the processing time. First, relations between the number of blocks and the processing time were investigated for problems with 50 and 200 variables and the result is shown in Figure 6.2 and 6.3, where the variables were shared on the average among blocks and one were distributed to every n blocks about the fractions. From these figures, it is observed that the processing time for the Dantzig-Wolfe decomposition principle-based method decreases more rapidly than two other methods with increasing the number of blocks, while the method needs more time than the two other methods when the number of blocks is relatively small.

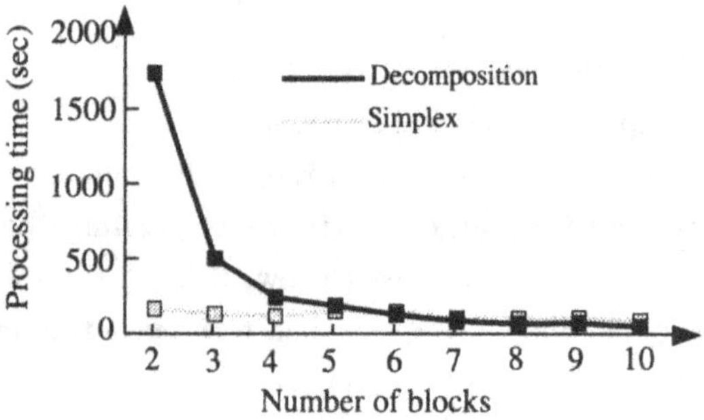

Figure 6.2. Relations between the number of blocks and the processing time (50 variables).

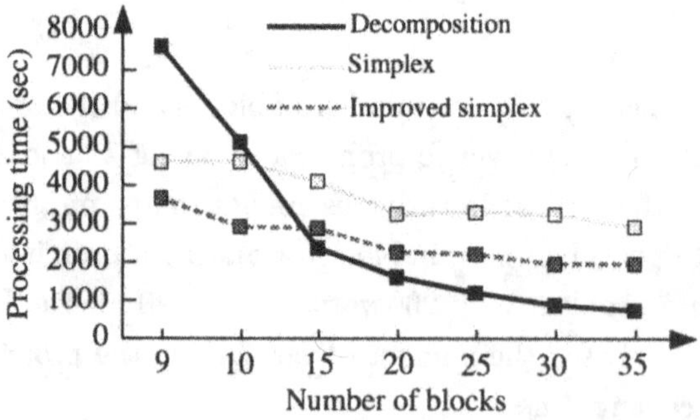

Figure 6.3. Relations between the number of blocks and the processing time (200 variables).

Next, ratios of the processing time for the simplex-based method to that for the Dantzig-Wolfe decomposition principle-based method were calculated and are shown in Figure 6.4 and 6.5. These figures show that the decomposition-based method becomes more efficient than the simplex-based method as the scale of the problem, the number of variables in the problem, or both become larger.

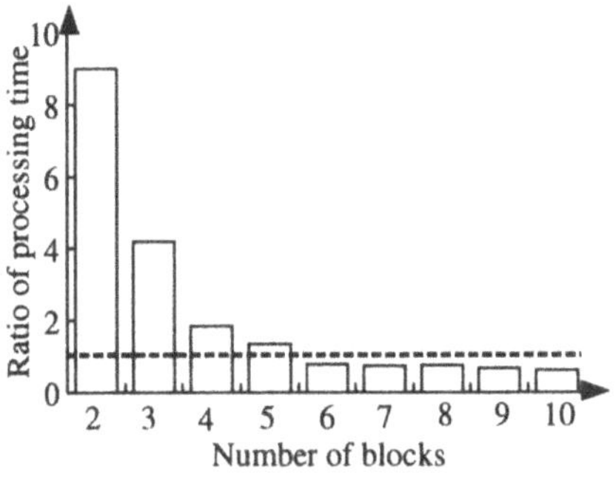

Figure 6.4. Ratios of the processing time for the simplex-based method to that for the decomposition-based method (50 variables).

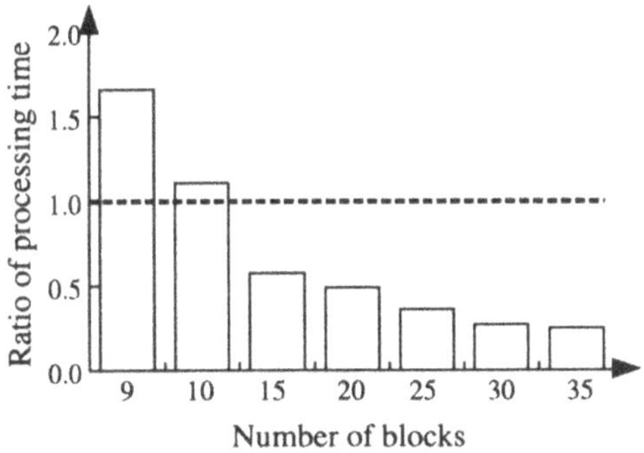

Figure 6.5. Ratios of the processing time for the simplex-based method to that for the decomposition-based method (200 variables).

Up to now, problems in which variables are shared on the average among blocks have been used as numerical examples. Here, consider a case that a certain block are allocated prominently many variables. Relations between the number of variables in the prominent block and the processing time are shown in Figure 6.6. From this figure, it is understood that if variables are shared on the average among blocks, the decomposition-based method is effective, but if there exists a block with remarkably many variables, it becomes inefficient.

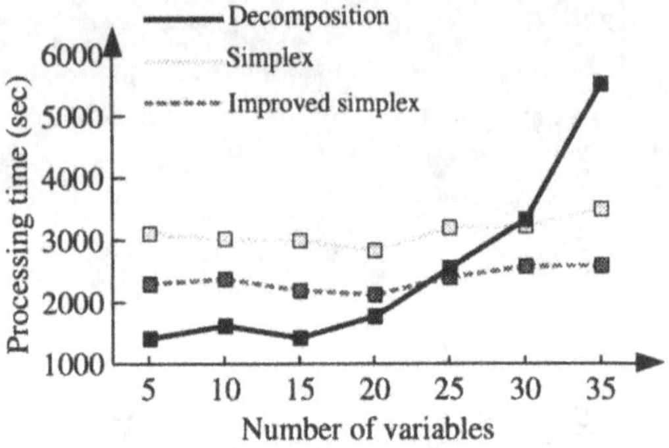

Figure 6.6. Relations between the number of variables in the prominent block and the processing time (200 variables and 20 blocks).

Required memory capacity. Required memory capacities depend on the size of the basis matrix mainly, so they are roughly estimated as follows:

- The decomposition-based method:

$$(m_0 + \underline{p} + M)^2$$

- The simplex-based method:

$$\left(m_0 + \underline{\sum_{i=1}^{p} m_i} + M \right)^2$$

where

m_0 : the number of coupling constraints
p : the number of blocks
m_i : the number of variables in the ith block
M : the number of constraints generated from objective functions

For example, required memory capacities in solving a problem with 2000 variables, 20 blocks, 3 objectives, 50 coupling constraints, and 50 constraints in every block can be calculated as follows:

- The decomposition-based method

$$(50 + 20 + 2)^2 \times 8 = 41472 \text{ (byte)}$$

- The simplex-based method

$$(50 + 20 \times 50 + 2)^2 \times 8 = 8853632 \text{ (byte)}$$

Relations between the number of blocks and the required memory capacity for problems with 2000 variables are shown in Figure 6.7. In the decomposition-based method, since the size of the basis matrix depends on the number of coupling constraints plus the number of blocks, the decomposition-based method was fairly more effective than the simplex method when the number of blocks was small.

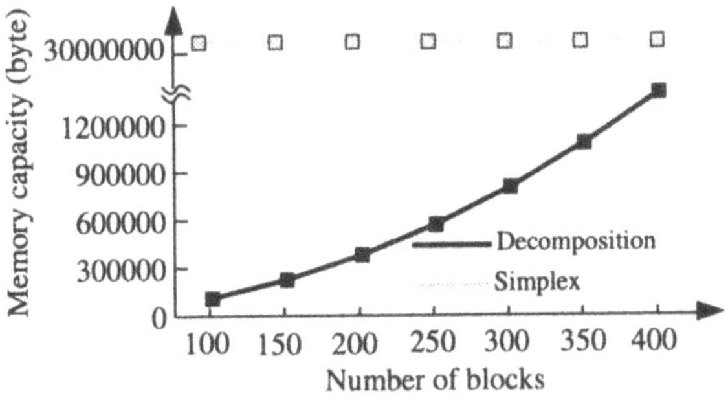

Figure 6.7. Relations between the number of blocks and the required memory capacity for problems with 2000 variables.

In our experimental environment, for a problem with 2000 variables, 3 objectives, 15 coupling constraints, 50 blocks, 40 constraints and 50 variables in every block, the decomposition-based method could be executed, but the simplex based method could not.

6.4 Conclusion

In this chapter, we have introduced not only interactive programming but also interactive fuzzy programming methods for multiobjective linear programming problems with block angular structures involving fuzzy numbers. In the proposed interactive decision making method, for the degree α and the reference levels specified by the DM, the corresponding α-Pareto optimal solution can be easily obtained by solving the minimax problems through the Dantzig-Wolfe decomposition method. The satisficing solution for the DM can be derived efficiently from among the α-Pareto optimal solutions by updating the reference levels and/or the degree α based on the current values of the α-Pareto optimal solution together with the trade-off information between the objective functions and the degree α. An illustrative numerical example was provided to demonstrate the feasibility of the proposed method.

Furthermore, by considering the vague nature of the DM's judgments, the fuzzy goals of the DM for the objective functions were quantified by eliciting the corresponding membership functions including nonlinear ones. Through the introduction of an M-α-Pareto optimality as an extended Pareto optimality concept, in the proposed interactive decision making method, if the DM specifies the degree α and the reference membership levels, the corresponding M-α-Pareto optimal solution can be obtained by solving the minimax problems for which the Dantzig-Wolfe decomposition method is applicable. Then, we presented a linear programming-based interactive fuzzy satisficing method for deriving a satisficing solution for the DM efficiently from an M-α-Pareto optimal solution set, by updating the reference membership levels and/or the degree α based on the current values of the M-α-Pareto optimal solution together with the trade-off information between the membership functions and the degree α. Through our computational experiments on numerical examples with both 50 and 200 variables, it was found that the decomposition-based method generally met with better results than the simplex-based methods. Especially, the efficiency of the decomposition-based method increased sharply with the scale of the problem. However, if the number of variables in a block is rather large or there exists a block with prominently many variables, the efficiency of the decomposition-based method becomes weak. Here, we compared the decomposition-based method with the revised simplex method and the revised simplex method with slight improvement. In the future, the comparison with other methods such as interior point methods and/or the proposition of new decomposition methods, which compensates for weaknesses of the Dantzig-Wolfe decomposition method, will be desirable. Also applications of the proposed method as well as extensions to nonlinear case will be required.

7. Genetic Algorithms with Decomposition Procedures

This chapter presents a detailed treatment of genetic algorithms with decomposition procedures as developed for large scale 0-1 knapsack problems with block angular structures. Through the introduction of a triple string representation and the corresponding decoding algorithm, it is shown that a potential solution satisfying not only block constraints but also coupling constraints can be obtained for each individual. The chapter also includes several numerical experiments.

7.1 Introduction

In Chapter 3, the Dantzig-Wolfe decomposition method for large scale linear programming problems with block angular structures is explained in detail. On the basis of the Dantzig-Wolfe decomposition method, in Chapters 4, 5 and 6, we focused mainly on large scale multiobjective linear programming problems with block angular structures under fuzziness and presented fuzzy linear programming, fuzzy multiobjective linear programming, interactive multiobjective programming, and interactive fuzzy multiobjective linear programming to derive the satisficing solution for the decision maker (DM) efficiently from a (M-) (α-) Pareto optimal solution set.

It should be observed here that, after the publication of the Dantzig-Wolfe decomposition algorithm [13, 14], the subsequent works on large scale linear and nonlinear programming problems with block angular structures have been numerous [58]. Unfortunately, however, a generalization of the results along this line for dealing with discrete optimization problems with block angular structures is not yet established.

Recently, for dealing with multiobjective multidimensional 0-1 knapsack problems with block angular structures, Kato and Sakawa [46, 47, 51] presented genetic algorithms with decomposition procedures. Through the introduction of a triple string representation and the corresponding decoding algorithm, it is shown that a potential solution satisfying both the block constraints and coupling constraints can be obtained for each individual.

In this chapter, for convenience in our subsequent discussion, genetic algorithms with decomposition procedures for multidimensional 0-1 knapsack problems with the block angular structure proposed by Kato and Sakawa [46, 47, 51] are explained in detail together with several numerical experiments.

7.2 Multidimensional 0-1 knapsack problems with block angular structures

As a relatively simple 0-1 version of the linear programming problem with the block angular structure discussed in Chapter 3, consider the following large scale 0-1 programming problem of the block angular form:

$$\left.\begin{array}{lrlcll} \text{minimize} & \boldsymbol{c}\boldsymbol{x} = & \boldsymbol{c}_1\boldsymbol{x}_1 & + \cdots + & \boldsymbol{c}_p\boldsymbol{x}_p & \\ \text{subject to} & A\boldsymbol{x} = & A_1\boldsymbol{x}_1 & + \cdots + & A_p\boldsymbol{x}_p & \leq \boldsymbol{b}_0 \\ & & B_1\boldsymbol{x}_1 & & & \leq \boldsymbol{b}_1 \\ & & & \ddots & & \vdots \\ & & & & B_p\boldsymbol{x}_p & \leq \boldsymbol{b}_p \\ & & \multicolumn{4}{l}{\boldsymbol{x}_j \in \{0,1\}^{n_j},\ j = 1,\ldots,p} \end{array}\right\} \quad (7.1)$$

where $\boldsymbol{c}_j$, $j = 1,\ldots,p$, are n_j dimensional cost factor row vectors, $\boldsymbol{x}_j$, $j = 1,\ldots,p$, are n_j dimensional column vectors of 0-1 decision variables, $A\boldsymbol{x} = A_1\boldsymbol{x}_1 + \cdots + A_p\boldsymbol{x}_p \leq \boldsymbol{b}_0$ denotes m_0 dimensional coupling constraints, and $A_j, j = 1,\ldots,p$, are $m_0 \times n_j$ coefficient matrices. The inequalities $B_j\boldsymbol{x}_j \leq \boldsymbol{b}_j$, $j = 1,\ldots,p$, are m_j dimensional block constraints, where B_j, $j = 1,\ldots,p$, are $m_j \times n_j$ coefficient matrices. Here, it is assumed that each element of A_j, B_j and $\boldsymbol{b}_j$ is nonnegative respectively. Then the problem (7.1) can be viewed as a multidimensional 0-1 knapsack problem with a block angular structure.

For example, consider a project selection problem in a company having a number of divisions, where each division has its own limited amounts of internal resources and the divisions are coupled by limited amounts of shared resources. The manager is to determine the projects to be actually approved so as to maximize the total profit under the resource constraints. Such a project selection problem can be formulated as a multidimensional 0-1 knapsack problem with a block angular structure expressed by (7.1).

7.3 Genetic algorithms with decomposition procedures

7.3.1 Coding and decoding

As outlined in Chapter 2, for 0-1 programming problems of the knapsack type, Sakawa et al. [82, 96, 97, 98, 106, 107, 109] proposed a double string representation as shown in Figure 7.1 and the corresponding decoding algorithm to generate only feasible solutions.

$\nu(1)$	$\nu(2)$	$\cdots$	$\nu(n)$
$g_{\nu(1)}$	$g_{\nu(2)}$	$\cdots$	$g_{\nu(n)}$

Figure 7.1. Double string.

In Figure 7.1, for a certain k, $\nu(k) \in \{1, \ldots, n\}$ denotes an index of a variable in the solution space, while $g_{\nu(k)}$, $k = 1, \ldots, n$ denotes the value (0 or 1) of the $\nu(k)$th variable.

In view of the special structure of the problem (7.1), it seems to be quite reasonable to define an individual $\mathbf{S}$ as an aggregation of p subindividuals $\mathbf{s}^j$, $j = 1, \ldots, p$, corresponding to the block constraint $B_j \boldsymbol{x}_j \leq \boldsymbol{b}_j$ as shown in Figure 7.2.

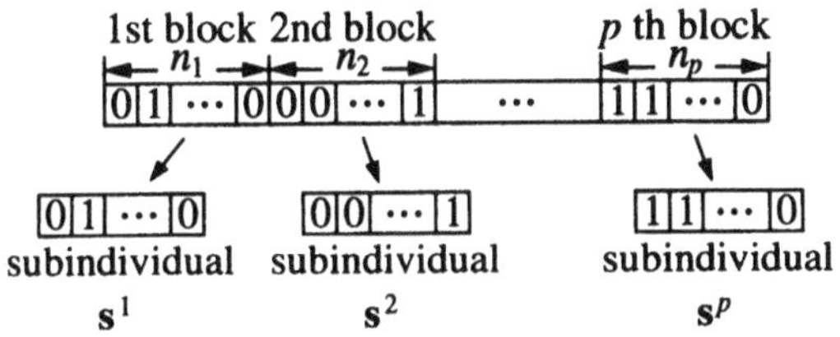

Figure 7.2. Division of a individual into p subindividuals.

If these subindividuals are represented by double strings, for each of the subindividuals $\mathbf{s}^j$, $j = 1, \ldots, p$, a phenotype (subsolution) satisfying each of the block constraints can be obtained by the decoding algorithm proposed by Sakawa et al. [82, 96, 97, 98, 106, 107, 109].

Unfortunately, however, the simple combination of these subsolutions does not always satisfy the coupling constraints. To cope with this problem, Kato and Sakawa [46, 47, 51] proposed a triple string representation as shown in Figure 7.3 and the corresponding decoding algorithm, as an extension of the double string representation and the corresponding decoding algorithm. By using the proposed representation and decoding algorithm, a phenotype (solution) satisfying both the block constraints and coupling constraints can be obtained for each individual $\mathbf{S} = (\mathbf{s}^1, \mathbf{s}^2, \ldots, \mathbf{s}^p)$. To be more explicit, in a

$\mathbf{s}^j =$

r^j			
$\nu^j(1)$	$\nu^j(2)$	$\cdots$	$\nu^j(n_j)$
$g^j_{\nu^j(1)}$	$g^j_{\nu^j(2)}$	$\cdots$	$g^j_{\nu^j(n_j)}$

Figure 7.3. Triple string representation.

triple string which represents a subindividual corresponding to the jth block, $r^j \in \{1, \ldots, p\}$ represents the priority of the jth block, $\nu^j(k) \in \{1, \ldots, n_j\}$ denotes an index of a variable in phenotype and $g^j_{\nu^j(k)}$ is a 0-1 value variable.

Decoding this individual (genotype) by means of the following algorithm, the resulting solution (phenotype) becomes always feasible. In the algorithm, n_j denotes the number of variables in the jth block, $j = 1, \ldots, p$, $\boldsymbol{\alpha}^j_{\nu^j(k)}$ is the $\nu^j(k)$th column vector in the jth coupling constraint coefficient matrix A_j, and $\boldsymbol{\beta}^j_{\nu^j(k)}$ is the $\nu^j(k)$th column vector in the jth block constraint coefficient matrix B_j.

Decoding algorithm for triple string

Step 1: Set $i = 1$, $\boldsymbol{\Sigma} = \mathbf{0}$ and proceed to Step 2.

Step 2: Find out such a block as $i = r^j$, and proceed to Step 3.

Step 3: For the above block j, set $k = 1$ and $\boldsymbol{\sigma} = \mathbf{0}$, and repeat the following procedures.

(a) If $g^j_{\nu^j(k)} = 1$, set $k = k+1$ and go to (b). Otherwise, i.e., if $g^j_{\nu^j(k)} = 0$, set $x^j_{\nu^j(k)} = 0$, $k = k+1$ and go to (c).

(b) If $\boldsymbol{\Sigma} + \boldsymbol{\alpha}^j_{\nu^j(k)} \leq \boldsymbol{b}_0$ and $\boldsymbol{\sigma} + \boldsymbol{\beta}^j_{\nu^j(k)} \leq \boldsymbol{b}_j$, set $x^j_{\nu^j(k)} = 1$, $\boldsymbol{\Sigma} = \boldsymbol{\Sigma} + \boldsymbol{\alpha}^j_{\nu^j(k)}$, $\boldsymbol{\sigma} = \boldsymbol{\sigma} + \boldsymbol{\beta}^j_{\nu^j(k)}$, and go to (c). Otherwise, set $x^j_{\nu^j(k)} = 0$ and go to (c).

(c) If $k > n_j$, go to Step 4 and regard $\boldsymbol{x}^j = (x^j_1, \ldots, x^j_{n_j})^T$ as phenotype of a subindividual $\mathbf{s}^j$ represented by triple string. Otherwise, return to (a).

Step 4: If $i = p$, stop and regard $\boldsymbol{x} = (\boldsymbol{x}^1, \ldots, \boldsymbol{x}^p)^T$ as phenotype of an individual $\mathbf{S}$. Otherwise, set $i = i + 1$ and return to Step 2.

To illustrate how the decoding algorithm actually works, consider a simple numerical example with 2 blocks and 5 variables as shown in Figure 7.4. Since the values in the upper string denote the priority of decoding, the second subindividual will be decoded first. Furthermore, since the first element of the middle string and the lower string of the second subindividual are '2' and '1' respectively (marked '_'), the second variable in the second block x_{22} is set to 1 if both the block constraints and coupling constraints are satisfied. In this example, all constraints are satisfied, and x_{22} is fixed to 1. Next, although the second element of the middle string and the lower string of the second subindividual are '1' and '1' respectively (marked '_'), the first variable of the second individual x_{21} is set to 0 since the block constraint is violated if $x_{21} = 1$. These procedures are repeated until all values of the variables in the phenotype are fixed. In this example, a feasible solution $[0, 0, 1, 0, 1]$ is obtained from the given individual by use of the decoding algorithm.

Problem (2 blocks, 5 variables)

$$\begin{array}{rcl} 3x_{11} + 4x_{12} + 3x_{13} + 2x_{21} + 6x_{22} & \leq & 9 \\ 2x_{11} + 5x_{12} + 9x_{13} & \leq & 15 \\ 7x_{21} + 5x_{22} & \leq & 10 \end{array}$$

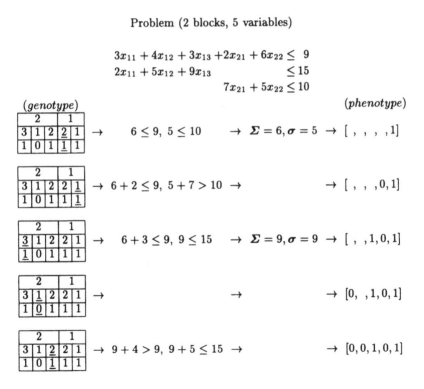

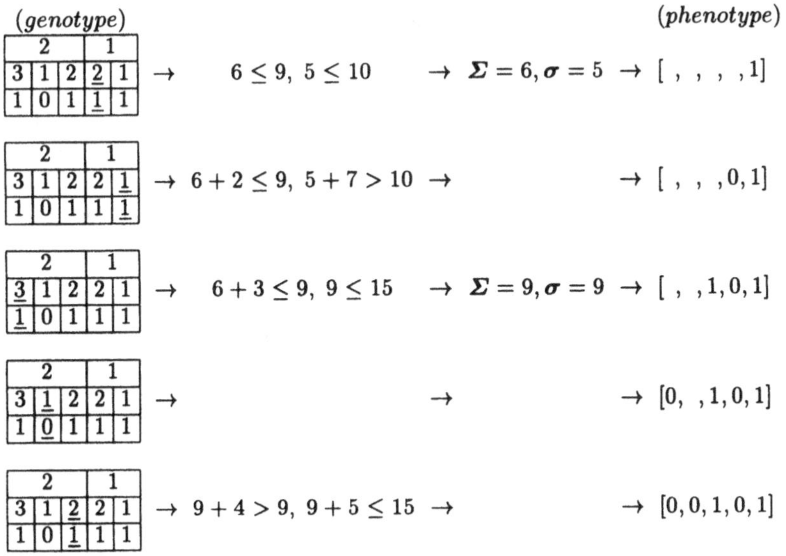

Figure 7.4. An example for the decoding algorithm.

7.3.2 Fitness and scaling

In genetic algorithms, an individual **S** is evaluated using some measure of fitness. As fitness of each individual **S** for the problem (7.1), it would be reasonable to adopt the following function

$$f(\mathbf{S}) = \frac{cx - \sum_{(i,j)\in I_+} c_{ij}}{\sum_{(i,j)\in I_-} c_{ij} - \sum_{(i,j)\in I_+} c_{ij}}, \tag{7.2}$$

where **S** denotes an individual represented by a triple string and x is the phenotype of **S**. Furthermore, $I_+ = \{(i,j) \mid c_{ij} \geq 0, 1 \leq i \leq n_j, 1 \leq j \leq p\}$ and $I_- = \{(i,j) \mid c_{ij} < 0, 1 \leq i \leq n_j, 1 \leq j \leq p\}$.

It should be noted here that the fitness $f(\mathbf{S})$ becomes as

$$f(\mathbf{S}) = \begin{cases} 0, & \text{if } x_{ij} = 1,\ (i,j) \in I_+ \text{ and } x_{ij} = 0,\ (i,j) \in I_- \\ 1, & \text{if } x_{ij} = 0,\ (i,j) \in I_+ \text{ and } x_{ij} = 1,\ (i,j) \in I_- \end{cases}$$

and the fitness $f(\mathbf{S})$ satisfies $0 \leq f(\mathbf{S}) \leq 1$. For convenience, fitness of an individual **S** is used as fitness of each subindividual $\mathbf{s}^j$, $j = 1, \ldots, p$, as in [46, 47, 51].

In the reproduction operation based on the ratio of fitness of each individual to the total fitness such as the expected value selection, it is a problem that the probability of selection depends on the relative ratio of fitness of each individual. Thus, the following linear scaling is adopted as in [46, 47, 51].

Linear scaling: Fitness $f_i = f(\mathbf{S}_i)$, of an ith individual $\mathbf{S}_i$ is transformed into f'_i as follows:

$$f'_i = a \cdot f_i + b,$$

where the coefficients a and b are determined so that the mean fitness f_{mean} of the population becomes a fixed point and the maximal fitness f_{max} of the population becomes twice as large as the mean fitness, i.e., $f_{\text{mean}} = a \cdot f_{\text{mean}} + b$ and $2 \cdot f_{\text{mean}} = a \cdot f_{\text{max}} + b$.

7.3.3 Reproduction

Various kinds of reproduction methods have been proposed. Among them, Sakawa et al. [82, 98] investigated the performance of each of six reproduction operators, i.e., ranking selection, elitist ranking selection, expected value selection, elitist expected value selection, roulette wheel selection and elitist roulette wheel selection, and as a result confirmed that elitist expected value selection is relatively efficient for multiobjective 0-1 programming problems incorporating the fuzzy goals of the decision maker. Based mainly on our experience [82], as a reproduction operator, elitist expected value selection – elitism and expected value selection combined together – is adopted, where elitism and expected value selection are summarized as follows:

Elitism: If the fitness of an individual in the past populations is larger than that of every individual in the current population, preserve this string into the current generation.

Expected value selection: For a population consisting of N individuals, the expected number of each subindividual of the nth individual $\mathbf{s}_n^j$, $j = 1, \ldots, p$, in the next population is given by

$$N_n = \frac{f(\mathbf{S}_n)}{\sum_{n=1}^{N} f(\mathbf{S}_n)} \times N.$$

Then, the integral part of N_n $(= [N_n])$ denotes the definite number of individual $\mathbf{s}_n^j$ preserved in the next population. While, using the decimal part of N_n $(= N_n - [N_n])$, the probability to preserve $\mathbf{s}_n^j$, $j = 1, \ldots, p$, in the next population is determined by

$$\frac{N_n - [N_n]}{\sum_{n=1}^{N} (N_n - [N_n])}.$$

7.3.4 Crossover

If a single-point crossover or multi-point crossover is directly applied to upper or middle string of individuals of triple string type, the kth element of the string of an offspring may take the same number that the k'th element takes. The same violation occurs in solving the traveling salesman problems or scheduling problems through genetic algorithms. In order to avoid this violation, a crossover method called partially matched crossover (PMX) is modified to be suitable for triple strings [46, 47, 51]. PMX is applied as usual for upper strings, whereas, for a couple of middle string and lower string, PMX for double strings [82, 96, 97, 98, 106, 107, 108, 109] is applied to every subindividual.

It is now appropriate to present the detailed procedures of the crossover method for triple strings.

Partially Matched Crossover (PMX) for upper string

Let

$$X = \boxed{r_X^1}\boxed{r_X^2}\boxed{\;\cdots\;}\boxed{r_X^p}$$

be the upper string of an individual and

$$Y = \boxed{r_Y^1}\boxed{r_Y^2}\boxed{\;\cdots\;}\boxed{r_Y^p}$$

be the upper string of another individual. Prepare copies X' and Y' of X and Y, respectively.

Step 1: Choose two crossover points at random on these strings, say, h and k $(h < k)$.

Step 2: Set $i = h$ and repeat the following procedures.

(a) Find j such that $r_{X'}^j = r_Y^i$. Then, interchange $r_{X'}^i$ with $r_{X'}^j$ and set $i = i + 1$.

(b) If $i > k$, stop and let X' be the offspring of X. Otherwise, return to (a).

Step 2 is carried out for Y' in the same manner, as shown in Figure 7.5.

Partially Matched Crossover (PMX) for double string

Let

$$X = \begin{array}{|c|c|c|c|}\hline \nu_X^j(1) & \nu_X^j(2) & \cdots & \nu_X^j(n_j) \\ \hline g^j_{\nu_X^j(1)} & g^j_{\nu_X^j(2)} & \cdots & g^j_{\nu_X^j(n_j)} \\ \hline \end{array}$$

be the middle and lower part of a subindividual in the jth subpopulation, and

$$Y = \begin{array}{|c|c|c|c|}\hline \nu_Y^j(1) & \nu_Y^j(2) & \cdots & \nu_Y^j(n_j) \\ \hline g^j_{\nu_Y^j(1)} & g^j_{\nu_Y^j(2)} & \cdots & g^j_{\nu_Y^j(n_j)} \\ \hline \end{array}$$

be the middle and lower part of another subindividual in the jth subpopulation. First, prepare copies X' and Y' of X and Y, respectively.

Step 1: Choose two crossover points at random on these strings, say, h and k $(h < k)$.

Step 2: Set $i = h$ and repeat the following procedures.

(a) Find i' such that $\nu_{X'}^j(i') = \nu_Y^j(i)$. Then, interchange $(\nu_{X'}^j(i),\ g^j_{\nu_{X'}^j(i)})^T$ with $(\nu_{X'}^j(i'),\ g^j_{\nu_{X'}^j(i')})^T$ and set $i = i + 1$.

(b) If $i > k$, stop. Otherwise, return to (a).

Step 3: Replace the part from h to k of X' with that of Y and let X' be the offspring of X.

This procedure is carried out for Y' and X in the same manner, as shown in Figure 7.6.

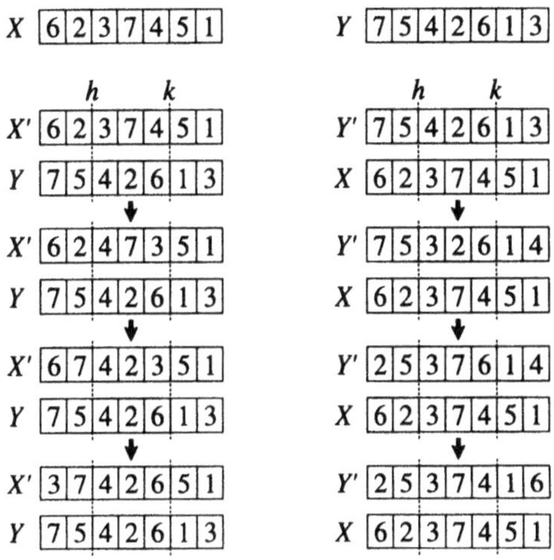

Figure 7.5. An example of PMX for upper string.

7.3.5 Mutation

It is considered that mutation plays the role of local random search in genetic algorithms. In [46, 47, 51], only for the lower string of a triple string, mutation of bit-reverse type is adopted and applied to every subindividual.

In [46, 47, 51], for the upper string and for the middle and lower string of the triple string, inversion defined by the following algorithm is adopted:

Step 1: After determining two inversion points h and k ($h < k$), pick out the part of the string from h to k.

Step 2: Arrange the substring in reverse order.

Step 3: Put the arranged substring back in the string.

Figure 7.7 illustrates examples of mutation.

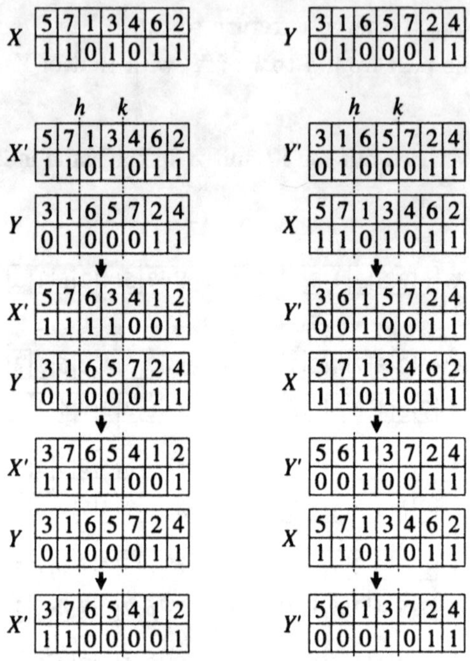

Figure 7.6. An example of PMX for double string.

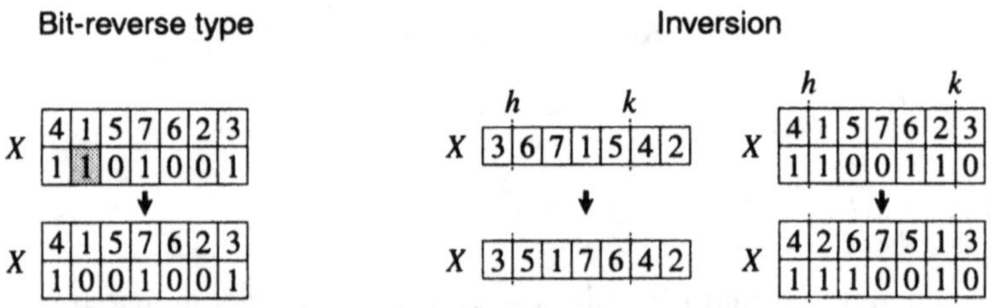

Figure 7.7. Examples of mutation.

7.3.6 Genetic algorithms with decomposition procedures

When applying genetic algorithms with decomposition procedures to the problem (7.1), an approximate optimal solution of desirable precision must be obtained in a proper time. For this reason, two parameters $I_{\min}$, which denotes the minimal search generation, and $I_{\max}$, which denotes the maximal search generation, are introduced in genetic algorithms with decomposition procedures.

Now we are ready to introduce the genetic algorithm with decomposition procedures as an approximate solution method for multidimensional 0-1 knapsack problems with block angular structures.

Genetic algorithm with decomposition procedures

Step 1: Set an iteration index (generation) $t = 0$ and determine the parameter values for the population size N, the probability of crossover p_c, the probability of mutation p_m, the probability of inversion p_i, the convergence criterion ε, the minimal search generation $I_{\min}$ and the maximal search generation $I_{\max}$.

Step 2: Generate N individuals whose subindividuals are of the triple string type at random.

Step 3: Evaluate each individual (subindividual) on the basis of phenotype obtained by the decoding algorithm and calculate the mean fitness f_{mean} and the maximal fitness $f_{\max}$ of the population. If $t > I_{\min}$ and $(f_{\max} - f_{\text{mean}})/f_{\max} < \varepsilon$, or, if $t > I_{\max}$, regard an individual with the maximal fitness as an optimal individual and terminate this program. Otherwise, set $t = t + 1$ and proceed to Step 4.

Step 4†: Apply the reproduction operator to all subpopulations $\{\mathbf{s}_n^j \mid n = 1, \ldots, N\}$, $j = 1, \ldots, p$.

Step 5†: Apply the PMX for double strings to the middle and lower part of every subindividual according to the probability of crossover p_c.

Step 6†: Apply the mutation operator of the bit-reverse type to the lower part of every subindividual according to the probability of mutation p_m, and apply the inversion operator for the middle and lower part of every subindividual according to the probability of inversion p_i.

Step 7: Apply the PMX for upper strings according to p_c.

Step 8: Apply the inversion operator for upper strings according to p_i and return to Step 3.

It should be noted here that, in the algorithm, the operations in the steps marked with † can be applied to every subindividual of all individuals independently. As a result, it is theoretically possible to reduce the amount of working memory needed to solve the problem and carry out parallel processing.

7.4 Numerical experiments

For investigating the feasibility and efficiency of the proposed method, consider multidimensional 0-1 knapsack problems with block angular structures having 30, 50, 70, 100, 150 and 200 variables.

Each element of $\boldsymbol{c}_j$, A_j and B_j, $j = 1, \ldots, p$, in the numerical example corresponding to the problem (7.1) was selected at random from the closed intervals $[-999, 0]$, $[0, 999]$ and $[0, 999]$, respectively. On the basis of these values, each element $b_j(k)$ of $\boldsymbol{b}_j$, $j = 0, \ldots, p$; $k = 1, \ldots, m_j$, was determined by

$$b_j(k) = \begin{cases} \gamma \times \sum_{j=1}^{p} \sum_{j'=1}^{n_j} A^j_{kj'}, & j = 0 \\ \delta \times \sum_{j'=1}^{n_j} B^j_{kj'}, & j = 1, \ldots, p \end{cases}$$

where positive constants γ and δ denote the degree of strictness of the coupling constraints and the block constraints respectively, and $A^j_{kj'}$'s and $B^j_{kj'}$'s denote elements of matrices A_j and B_j respectively.

Our numerical experiments were performed on a personal computer (processor: Celeron 466MHz, memory: 128MB, OS: Windows NT 4.0) using a Visual C++ compiler (version 6.0).

For comparison, the genetic algorithm with double strings [82, 96, 97, 98, 106, 107, 109] outlined in Subsection 2.7.2 was directly applied to the same problems. Also, in order to compare the obtained results with the corresponding exact optimal solutions or incumbent values, the same problems were solved using LP_SOLVE [5] by M. Berkelaar[†].

The parameter values used in both the genetic algorithm with double strings and genetic algorithm with decomposition procedures were set as follows: population size = 100, probability of crossover $p_c = 0.8$, probability of mutation $p_m = 0.01$, probability of inversion $p_i = 0.03$, $\varepsilon = 0.01$, $I_{\max} = 500$ and $I_{\min} = 100$. Observe that these parameter values were found through our experiences and these values used all of the trials of both the genetic algorithm with double strings and genetic algorithm with decomposition procedures.

First consider the following multidimensional 0-1 knapsack problem with 3 blocks, 30 (10+12+8) variables, 6 coupling constraints, and 19 (6+8+5)

[†] LP_SOLVE [5] solves (mixed integer) linear programming problems. The implementation of the simplex kernel was mainly based on the text by Orchard-Hays [69]. The mixed integer branch and bound part was inspired by Dakin [11].

block constraints:

$$\left.\begin{array}{rl} \text{minimize} & \boldsymbol{c}_1\boldsymbol{x}_1 + \boldsymbol{c}_2\boldsymbol{x}_2 + \boldsymbol{c}_3\boldsymbol{x}_3 \\ \text{subject to} & A_1\boldsymbol{x}_1 + A_2\boldsymbol{x}_2 + A_3\boldsymbol{x}_3 \le \boldsymbol{b}_0 \\ & B_1\boldsymbol{x}_1 \le \boldsymbol{b}_1 \\ & B_2\boldsymbol{x}_2 \le \boldsymbol{b}_2 \\ & B_3\boldsymbol{x}_3 \le \boldsymbol{b}_3 \\ & \boldsymbol{x}_j \in \{0,1\}^{n_j},\ j = 1,2,3. \end{array}\right\} \tag{7.3}$$

As a numerical example with 30 variables, we used the following values of $\boldsymbol{c} = (\boldsymbol{c}_1, \boldsymbol{c}_2, \boldsymbol{c}_3)$, A_j and B_j, $j = 1,2,3$, which were determined at random from $[-999, 0]$, $[0, 999]$ and $[0, 999]$, respectively.

$$\begin{array}{llllllllll} \boldsymbol{c} = (& -6 & -292 & -997 & -260 & -638 & -591 & -194 & -339 & -521 & -177 \\ & -943 & -927 & -167 & -266 & -843 & -999 & -830 & -41 & -996 & -779 \\ & -755 & -540 & -488 & -922 & -736 & -589 & -696 & -213 & -154 & -334\ ) \end{array}$$

$$A_1 = \begin{pmatrix} 836 & 560 & 132 & 381 & 652 & 546 & 982 & 95 & 646 & 351 \\ 124 & 814 & 302 & 207 & 757 & 695 & 468 & 835 & 857 & 385 \\ 930 & 181 & 146 & 954 & 347 & 450 & 607 & 308 & 477 & 280 \\ 789 & 521 & 119 & 836 & 973 & 743 & 248 & 702 & 687 & 447 \\ 464 & 730 & 910 & 837 & 961 & 454 & 138 & 830 & 316 & 284 \\ 360 & 841 & 119 & 743 & 139 & 738 & 325 & 679 & 959 & 516 \end{pmatrix}$$

$$A_2 = \begin{pmatrix} 585 & 80 & 780 & 298 & 868 & 364 & 21 & 815 & 39 & 337 & 667 & 104 \\ 613 & 211 & 927 & 681 & 923 & 546 & 537 & 747 & 775 & 609 & 199 & 126 \\ 55 & 332 & 725 & 278 & 719 & 716 & 414 & 803 & 37 & 78 & 367 & 841 \\ 244 & 440 & 150 & 467 & 644 & 818 & 413 & 79 & 823 & 603 & 12 & 676 \\ 287 & 499 & 862 & 280 & 495 & 975 & 61 & 5 & 742 & 817 & 253 & 887 \\ 891 & 423 & 976 & 4 & 959 & 171 & 228 & 699 & 701 & 915 & 722 & 138 \end{pmatrix}$$

$$A_3 = \begin{pmatrix} 721 & 962 & 839 & 275 & 654 & 290 & 58 & 251 \\ 824 & 83 & 391 & 274 & 517 & 668 & 401 & 662 \\ 177 & 57 & 764 & 71 & 948 & 177 & 666 & 914 \\ 92 & 880 & 120 & 938 & 318 & 831 & 356 & 28 \\ 886 & 514 & 275 & 0 & 457 & 733 & 846 & 531 \\ 339 & 4 & 865 & 225 & 274 & 379 & 898 & 560 \end{pmatrix}$$

$$B_1 = \begin{pmatrix} 239 & 745 & 488 & 287 & 871 & 484 & 131 & 506 & 611 & 563 \\ 532 & 760 & 347 & 252 & 606 & 659 & 692 & 891 & 266 & 137 \\ 320 & 533 & 649 & 402 & 245 & 708 & 938 & 95 & 475 & 105 \\ 436 & 695 & 921 & 455 & 47 & 692 & 990 & 390 & 824 & 582 \\ 750 & 564 & 735 & 180 & 86 & 810 & 307 & 319 & 750 & 623 \\ 306 & 442 & 702 & 751 & 99 & 126 & 669 & 159 & 873 & 68 \end{pmatrix}$$

$$B_2 = \begin{pmatrix} 298 & 169 & 342 & 569 & 847 & 46 & 126 & 39 & 778 & 806 & 435 & 561 \\ 203 & 544 & 227 & 45 & 544 & 43 & 494 & 993 & 352 & 959 & 828 & 427 \\ 667 & 366 & 933 & 986 & 837 & 923 & 220 & 745 & 194 & 471 & 170 & 772 \\ 501 & 312 & 595 & 966 & 569 & 983 & 463 & 197 & 964 & 774 & 449 & 284 \\ 186 & 802 & 860 & 848 & 249 & 893 & 95 & 558 & 63 & 80 & 289 & 811 \\ 510 & 975 & 310 & 380 & 644 & 815 & 757 & 47 & 900 & 453 & 107 & 769 \\ 727 & 75 & 689 & 396 & 559 & 929 & 623 & 481 & 510 & 846 & 383 & 206 \\ 725 & 830 & 147 & 958 & 769 & 504 & 588 & 589 & 977 & 692 & 257 & 888 \end{pmatrix}$$

$$B_3 = \begin{pmatrix} 162 & 913 & 217 & 296 & 593 & 75 & 93 & 399 \\ 648 & 147 & 175 & 225 & 41 & 92 & 753 & 878 \\ 44 & 839 & 318 & 547 & 593 & 577 & 809 & 407 \\ 699 & 153 & 534 & 461 & 60 & 860 & 953 & 835 \\ 291 & 381 & 950 & 574 & 659 & 188 & 769 & 464 \end{pmatrix}$$

For $(\gamma, \delta) = (0.25, 0.25)$, $(0.50, 0.50)$, and $(0.75, 0.75)$, 10 trails to each example were performed through both the genetic algorithm with double strings (GADS) and genetic algorithm with decomposition procedures (GADP). Also, in comparison with exact optimal solutions, each example was solved by LP_SOLVE [5]. Table 7.1 shows the experimental results for $(\gamma, \delta) = (0.25, 0.25)$, $(0.50, 0.50)$, and $(0.75, 0.75)$, where Best, Average, Worst, Time, AG and # represent the best value, average value, worst value, average processing time, average generation for obtaining the best value, and the number of best solution in 10 trials, respectively.

For problems with 30 variables, it can be seen from Table 7.1 that optimal solutions were obtained on 10 times out of 10 trials for both GADS and GADP. However, concerning the processing time, as expected, LP_SOLVE was much faster than GADP and GADS. It should also be noted that GADP requires more processing time than GADS. As a result, for problems with 30 variables, there is no evidence that would reveal an advantage of GADP over GADS and LP_SOLVE.

Next, consider another multidimensional 0-1 knapsack problem with a block angular structure, which has 5 blocks, 50 (=10+12+8+13+7) variables and 10 coupling constraints. The results obtained through 10 times trails of GADS and GADP for each of the problems are shown in Table 7.2 together with the experimental results by LP_SOLVE. From Table 7.2, it can be seen that GADP succeeds 10 times out of 10 trials for $(\gamma, \delta) = (0.25, 0.25)$ and $(0.50, 0.50)$, while LP_SOLVE cannot locate an optimal solution for $(\gamma, \delta) = (0.50, 0.50)$. Furthermore, the required processing time of GADP is lower than that of GADS and LP_SOLVE. As a result, for problems with 50 variables, GADP seems to be more desirable than GADS and LP_SOLVE.

Table 7.1. Experimental results for 30 variables (10 trails)

(γ, δ)	Methods	Best	Average	Worst	Time (sec)	AG	#
(0.25, 0.25)	GADS	−4011	−4011.0	−4011	2.78	11.8	10
	GADP	−4011	−4011.0	−4011	3.02	10.8	10
	LP_SOLVE	−4011 (Optimum)			1.04	——	
(0.50, 0.50)	GADS	−9726	−9726.0	−9726	3.31	69.7	10
	GADP	−9726	−9726.0	−9726	3.49	47.2	10
	LP_SOLVE	−9726 (Optimum)			0.90	——	
(0.75, 0.75)	GADS	−12935	−12935.0	−12935	3.94	158.6	10
	GADP	−12935	−12935.0	−12935	4.01	81.8	10
	LP_SOLVE	−12935 (Optimum)			1.35	——	

Table 7.2. Experimental results for 50 variables (10 trails)

(γ, δ)	Methods	Best	Average	Worst	Time (sec)	AG	#
(0.25, 0.25)	GADS	−6848	−6848.0	−6848	6.89	51.8	10
	GADP	−6848	−6848.0	−6848	6.01	32.8	10
	LP_SOLVE	−6848 (Optimum)			1.64×10^1	——	
(0.50, 0.50)	GADS	−14947	−14890.5	−14784	8.70	316	2
	GADP	−14947	−14947.0	−14947	6.93	240	10
	LP_SOLVE	——			——	——	
(0.75, 0.75)	GADS	−20469	−20349.7	−20222	1.08×10^1	347.9	1
	GADP	−20469	−20459.9	−20378	7.97	200.6	9
	LP_SOLVE	−20469 (Optimum)			1.08×10^1	——	

Similar computational experiments were performed on numerical examples with 70, 100, 150 and 200 variables, and the corresponding results are shown in Tables 7.3, 7.4, 7.5 and 7.6, respectively.

It is significant to note here that LP_SOLVE cannot locate optimal solutions for problems with 70, 100, 150 and 200 variables, while LP_SOLVE occasionally gives incumbent solutions for problems with 70 and 100 variables. On the contrary, although the accuracy of the best solutions obtained through GADP and GADS tends to decrease if compared to the case of 30 variables or 50 variables, on the average GADP gives better results than GADS with respect to the accuracy of the obtained solutions. Furthermore, the processing times for GADS increase more rapidly than for GADP when increasing the number of variables. Therefore, it is interesting to see how the processing times of both GADS and GADP change with the size of the problem.

Table 7.3. Experimental results for 70 variables (10 trails)

(γ, δ)	Methods	Best	Average	Worst	Time (sec)	AG	#
	GADS	−10342	−10342.0	−10342	1.27×10^1	186.1	10
(0.25, 0.25)	GADP	−10342	−10342.0	−10342	9.26	153.5	10
	LP_SOLVE	−9256 (Incumbent)			9.10	——	
	GADS	−20763	−20554.7	−20329	1.61×10^1	338.3	1
(0.50, 0.50)	GADP	−21132	−20944.0	−20781	1.06×10^1	356.3	1
	LP_SOLVE	——			——	——	
	GADS	−29111	−28831.6	−28510	1.97×10^1	326.5	1
(0.75, 0.75)	GADP	−29653	−29307.7	−29110	1.22×10^1	348.0	1
	LP_SOLVE	−29639 (Incumbent)			2.25×10^1	——	

Table 7.4. Experimental results for 100 variables (10 trails)

(γ, δ)	Methods	Best	Average	Worst	Time (sec)	AG	#
	GADS	−13669	−13399.2	−13091	2.39×10^1	332.3	1
(0.25, 0.25)	GADP	−13654	−13463.8	−13318	1.41×10^1	350.3	1
	LP_SOLVE	——			——	——	
	GADS	−29012	−27622.8	−26910	3.01×10^1	305.1	1
(0.50, 0.50)	GADP	−28452	−27870.9	−27340	1.65×10^1	352.3	1
	LP_SOLVE	——			——	——	
	GADS	−39738	−38825.4	−37515	3.70×10^1	239.0	1
(0.75, 0.75)	GADP	−40586	−39741.5	−39284	1.87×10^1	361.8	1
	LP_SOLVE	−41538 (Incumbent)			2.59×10^1	——	

Table 7.5. Experimental results for 150 variables (10 trails)

(γ, δ)	Methods	Best	Average	Worst	Time (sec)	AG	#
	GADS	−19469	−18961.7	−18593	5.91×10^1	287.6	1
(0.25, 0.25)	GADP	−19452	−19280.7	−18765	2.45×10^1	255.1	1
	LP_SOLVE	——			——	——	
	GADS	−39332	−38931.8	−38485	8.00×10^1	310.7	1
(0.50, 0.50)	GADP	−41058	−40017.1	−39178	2.88×10^1	275.8	1
	LP_SOLVE	——			——	——	
	GADS	−56825	−55617.2	−54851	1.01×10^2	358.9	1
(0.75, 0.75)	GADP	−57888	−56775.0	−55911	3.34×10^1	372.2	1
	LP_SOLVE	——			——	——	

Table 7.6. Experimental results for 200 variables (10 trails)

(γ, δ)	Methods	Best	Average	Worst	Time (sec)	AG	#
	GADS	−25493	−24714.0	−24062	1.49×10^2	294.0	1
(0.25, 0.25)	GADP	−25407	−25016.5	−24553	3.65×10^1	292.8	1
	LP_SOLVE		——		——	——	
	GADS	−51636	−50046.4	−49052	2.03×10^2	310.7	1
(0.50, 0.50)	GADP	−51601	−51011.6	−50252	4.33×10^1	282.9	1
	LP_SOLVE		——		——	——	
	GADS	−73433	−72134.9	−71076	2.56×10^2	301.9	1
(0.75, 0.75)	GADP	−74434	−73452.9	−72374	5.05×10^1	326.1	1
	LP_SOLVE		——		——	——	

Figure 7.8 shows processing times for typical problems with 30, 50, 70, 100, 150 and 200 variables through GADP and GADS. From Figure 7.8, it is observed that the processing time, as a function of the problem size, increases in a linear fashion for GADP, while increasing in a quadratic fashion for GADS.

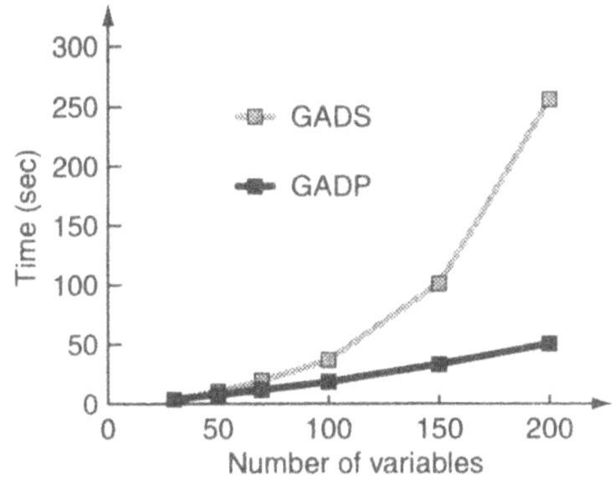

Figure 7.8. Processing time for typical problems with 30, 50, 70, 100, 150 and 200 variables (GADS versus GADP).

It is now appropriate to see how the processing time of GADP changes with the increased size of the problem. Figure 7.9 shows processing times for typical problems with 30, 50, 70, 100, 150, 200, 300, 500 and 1000 variables through GADP. As depicted in Figure 7.9, it can be seen that the processing time of GADP increases almost linearly with the size of the problem.

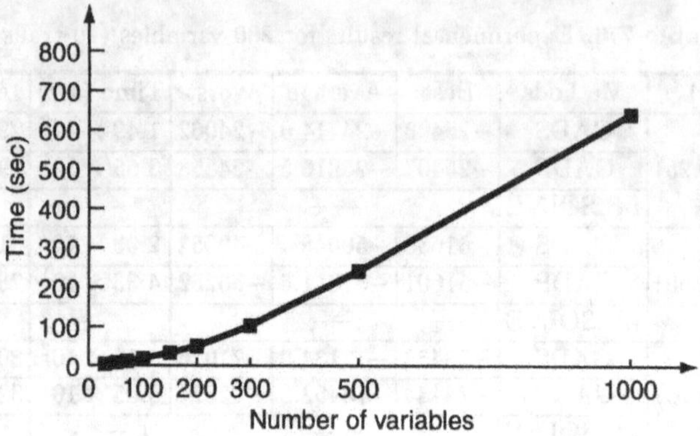

Figure 7.9. Processing time for typical problems with 30, 50, 70, 100, 150, 200, 300, 500 and 1000 variables (GADP).

From our numerical experiments through GADS and GADP, as an approximate solution method for multidimensional 0-1 knapsack problems with block angular structures, it is confirmed that GADP is more efficient and effective than GADS.

8. Large Scale Fuzzy Multiobjective 0-1 Programming

This chapter can be viewed as the multiobjective version of Chapter 7 and treats large scale multiobjective 0-1 knapsack problems with block angular structures by incorporating the fuzzy goals of the decision maker. On the basis of the genetic algorithm with decomposition procedures presented in Chapter 7, interactive fuzzy multiobjective 0-1 programming as well as fuzzy multiobjective 0-1 programming, both proposed by the authors, are explained in detail together with several numerical experiments.

8.1 Multiobjective multidimensional 0-1 knapsack problems with block angular structures

As the multiobjective version of Chapter 7, consider a large scale multiobjective multidimensional 0-1 knapsack problem of the following block angular form [46, 47]:

$$\left.\begin{array}{rl}\text{minimize} & \boldsymbol{c}_1\boldsymbol{x} = \boldsymbol{c}_{11}\boldsymbol{x}_1 + \cdots + \boldsymbol{c}_{1p}\boldsymbol{x}_p \\ \vdots & \qquad\qquad\qquad \vdots \\ \text{minimize} & \boldsymbol{c}_k\boldsymbol{x} = \boldsymbol{c}_{k1}\boldsymbol{x}_1 + \cdots + \boldsymbol{c}_{kp}\boldsymbol{x}_p \\ \text{subject to} & A\boldsymbol{x} = A_1\boldsymbol{x}_1 + \cdots + A_p\boldsymbol{x}_p \leq \boldsymbol{b}_0 \\ & \qquad\quad B_1\boldsymbol{x}_1 \qquad\qquad\quad \leq \boldsymbol{b}_1 \\ & \qquad\qquad\qquad \ddots \qquad\qquad \vdots \\ & \qquad\qquad\qquad\qquad B_p\boldsymbol{x}_p \leq \boldsymbol{b}_p \\ & \boldsymbol{x}_j \in \{0,1\}^{n_j}, \ j = 1, \ldots, p \end{array}\right\} \tag{8.1}$$

where $\boldsymbol{c}_{ij}$, $i = 1, \ldots, k$, $j = 1, \ldots, p$, are n_j dimensional row vectors, $\boldsymbol{x}_j$, $j = 1, \ldots, p$, are n_j dimensional column vectors of 0-1 decision variables, $A\boldsymbol{x} = A_1\boldsymbol{x}_1 + \cdots + A_p\boldsymbol{x}_p \leq \boldsymbol{b}_0$ denotes m_0 dimensional coupling constraints, A_j, $j = 1, \ldots, p$, are $m_0 \times n_j$ coefficient matrices. The inequalities $B_j\boldsymbol{x}_j \leq \boldsymbol{b}_j$,

$j = 1, \ldots, p$, are m_j dimensional block constraints with respect to $\boldsymbol{x}_j$, and, B_j, $j = 1, \ldots, p$, are $m_j \times n_j$ coefficient matrices. Here, it is assumed that each element of A_j, B_j and $\boldsymbol{b}_j$ is nonnegative respectively. Then the problem (8.1) can be viewed as a multiobjective multidimensional 0-1 knapsack problem with a block angular structure.

For example, consider a project selection problem in a company having a number of divisions, where each division has its own limited amounts of internal resources and the divisions are coupled by limited amounts of shared resources. The manager is to determine the projects to be actually approved so as to maximize the total profit and minimize the total amount of waste under the resource constraints. Such a project selection problem can be formulated as a multiobjective multidimensional 0-1 knapsack problem with a block angular structure expressed by (8.1).

As discussed thus far, for multiobjective programming problems to optimize multiple conflicting objective functions under the given constraints, the Pareto optimal solution is defined as a reasonable solution in place of a complete optimal solution which optimizes all objective functions [78].

8.2 Fuzzy goals

By considering the vague nature of human judgments for the large scale multiobjective multidimensional 0-1 knapsack problem (8.1), it is quite natural to assume that the decision maker (DM) may have a fuzzy goal such as "$\boldsymbol{c}_i\boldsymbol{x}$ should be substantially less than or equal to some value". Such a fuzzy goal of the DM can be quantified by eliciting the corresponding membership function. Here, for simplicity, the linear membership function

$$\mu_i(\boldsymbol{c}_i\boldsymbol{x}) = \begin{cases} 0, & \boldsymbol{c}_i\boldsymbol{x} > z_i^0 \\ \dfrac{\boldsymbol{c}_i\boldsymbol{x} - z_i^0}{z_i^1 - z_i^0}, & z_i^1 < \boldsymbol{c}_i\boldsymbol{x} \leq z_i^0 \\ 1, & \boldsymbol{c}_i\boldsymbol{x} \leq z_i^1 \end{cases} \tag{8.2}$$

is assumed for representing the fuzzy goal of the DM [78, 142, 145], where z_i^0 and z_i^1 denote the values of the objective function $\boldsymbol{c}_i\boldsymbol{x}$ whose degree of membership function are 0 and 1, respectively. These values are subjectively determined through an interaction with the DM. Figure 8.1 illustrates the graph of the possible shape of the linear membership function.

As one of the possible ways to help the DM determine z_i^0 and z_i^1, it is convenient to calculate the individual minimal value $z_i^{\min} = \min_{\boldsymbol{x} \in X} \boldsymbol{c}_i\boldsymbol{x}$

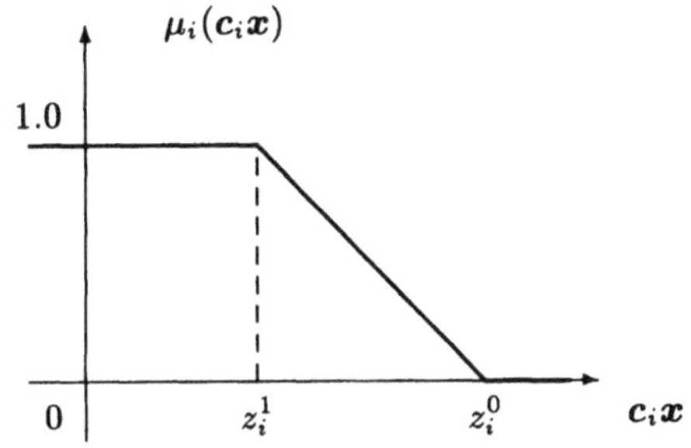

Figure 8.1. Linear membership function.

and maximal value $z_i^{\max} = \max_{x \in X} c_i x$ of each objective function under the given constrained set, where X is the constrained set of the large scale multiobjective multidimensional 0-1 knapsack problem (8.1). Then by taking account of the calculated individual minimum and maximum of each objective function, the DM is asked to assess z_i^0 and z_i^1 in the interval $[z_i^{\min}, z_i^{\max}]$, $i = 1, \ldots, k$.

Zimmermann [142] suggested a way to determine the linear membership function $\mu_i(c_i x)$, by assuming the existence of an optimal solution x^{io} of the individual objective function minimization problem under the constraints defined by

$$\min\{c_i x \mid x \in X\}, \ i = 1, \ldots, k. \tag{8.3}$$

To be more specific, using the individual minimum

$$z_i^{\min} = c_i x^{io} = \min\{c_i x \mid Ax \leq b, \ x \in \{0,1\}^n\}, \ i = 1, \ldots, k, \tag{8.4}$$

together with

$$z_i^{\mathrm{m}} = \max(c_i x^{1o}, \ldots, c_i x^{i-1,o}, c_i x^{i+1,o}, \ldots, c_i x^{ko}), \ i = 1, \ldots, k, \tag{8.5}$$

Zimmermann [142] determined the linear membership function as in (8.2) by choosing $z_i^1 = z_i^{\min}$ and $z_i^0 = z_i^{\mathrm{m}}$.

8.3 Fuzzy multiobjective 0-1 programming

8.3.1 Fuzzy programming through genetic algorithms

Having elicited the linear membership functions $\mu_i(c_i x)$, $i = 1, \ldots, k$, from the decision maker (DM) for each of the objective function $c_i x$, $i = 1, \ldots, k$, if we introduce a general aggregation function

$$\mu_D(\boldsymbol{x}) = \mu_D(\mu_1(\boldsymbol{c}_1\boldsymbol{x}), \ldots, \mu_k(\boldsymbol{c}_k\boldsymbol{x})), \tag{8.6}$$

a fuzzy multiobjective decision making problem can be defined by

$$\underset{\boldsymbol{x} \in X}{\text{maximize}} \ \mu_D(\boldsymbol{x}). \tag{8.7}$$

Observe that the value of the aggregation function $\mu_D(\boldsymbol{x})$ can be interpreted as representing an overall degree of satisfaction with the DM's multiple fuzzy goals [78]. As the aggregation function, if we adopt the fuzzy decision or minimum operator of Bellman and Zadeh, the large scale multiobjective multidimensional 0-1 knapsack problem (8.1) can be interpreted as

$$\left.\begin{array}{rl} \text{maximize} & \mu_D(\boldsymbol{x}) = \min_{i=1,\ldots,k} \left\{\mu_i(\boldsymbol{c}_i\boldsymbol{x})\right\} \\ \text{subject to} & A\boldsymbol{x} = A_1\boldsymbol{x}_1 + \cdots + A_p\boldsymbol{x}_p \leq \boldsymbol{b}_0 \\ & B_1\boldsymbol{x}_1 \qquad\qquad\qquad \leq \boldsymbol{b}_1 \\ & \qquad\qquad \ddots \qquad\qquad \vdots \\ & \qquad\qquad\qquad B_p\boldsymbol{x}_p \leq \boldsymbol{b}_p \\ & \boldsymbol{x}_j \in \{0,1\}^{n_j}, \quad j = 1, \ldots, p. \end{array}\right\} \tag{8.8}$$

The constraints of this problem preserve the block angular structure; genetic algorithms with decomposition procedures explained in Chapter 7 is applicable.

It should be noted here that the fitness of an individual $\mathbf{S}$ is defined as

$$f(\mathbf{S}) = \min_{i=1,\ldots,k} \left\{\mu_i(\boldsymbol{c}_i\boldsymbol{x})\right\} \tag{8.9}$$

where $\mathbf{S}$ and $\boldsymbol{x}$ respectively denote an individual represented by the triple string and the phenotype of $\mathbf{S}$. For convenience, fitness of an individual $\mathbf{S}$ is used as fitness of each subindividual $\mathbf{s}^j$, $j = 1, \ldots, p$.

We can now construct the algorithm in order to derive the satisficing solution for the DM to the large scale multiobjective multidimensional 0-1 knapsack problem (8.1) incorporating the fuzzy goals.

Fuzzy multiobjective 0-1 programming

Step 0: Calculate the individual minimum and maximum of each objective function under the given constraints.

Step 1*: Elicit a membership function from the DM for each of the objective function by taking account of the calculated individual minimum and maximum of each objective function.

Step 2: Set an iteration index (generation) $t = 0$ and determine the parameter values for the convergence criterion ε, the minimal search generation $I_{\min}$, the maximal search generation $I_{\max}$, the probability of crossover p_c, the probability of mutation p_m, the probability of inversion p_i.

Step 3: Generate N individuals of the triple string type at random as an initial population.

Step 4: Evaluate each individual (subindividual) on the basis of phenotype obtained by the proposed decoding algorithm and calculate the mean fitness f_{mean} and the maximal fitness $f_{\max}$ of the population. If $t > I_{\min}$ and $(f_{\max} - f_{\text{mean}})/f_{\max} < \varepsilon$, or, if $t > I_{\max}$, regard an individual with the maximal fitness as the best individual, and stop. Otherwise, set $t = t + 1$ and proceed to Step 5.

Step 5†: Apply the reproduction operator to every subindividual.

Step 6†: Apply the PMX for double strings to the middle and lower part of every subindividual according to the probability of crossover p_c.

Step 7†: Apply the mutation operator of the bit-reverse type to the lower part of every subindividual according to the probability of mutation p_m, and apply the inversion operator for the middle and lower part of every subindividual according to the probability of inversion p_i.

Step 8: Apply the PMX for upper strings according to p_c.

Step 9: Apply the inversion operator for upper strings according to p_i and return to Step 4.

In the algorithm, observe that the operations in the steps marked with † can be applied to every subindividual independently. Consequently, it is theoretically possible to reduce the amount of working memory and carry out parallel processing.

8.3.2 Numerical experiments

To illustrate the proposed method, consider the following multiobjective 0-1 programming problem with the block angular structure (3 objectives, 5 blocks, 50 (=10+12+8+13+7) variables, 10 coupling constraints, and 34 (7+8+5+10+4) block constraints):

$$\left.\begin{array}{ll}
\text{minimize} & \boldsymbol{c}_{11}\boldsymbol{x}_1 + \boldsymbol{c}_{12}\boldsymbol{x}_2 + \boldsymbol{c}_{13}\boldsymbol{x}_3 + \boldsymbol{c}_{14}\boldsymbol{x}_4 + \boldsymbol{c}_{15}\boldsymbol{x}_5 \\
\text{minimize} & \boldsymbol{c}_{21}\boldsymbol{x}_1 + \boldsymbol{c}_{22}\boldsymbol{x}_2 + \boldsymbol{c}_{23}\boldsymbol{x}_3 + \boldsymbol{c}_{24}\boldsymbol{x}_4 + \boldsymbol{c}_{25}\boldsymbol{x}_5 \\
\text{minimize} & \boldsymbol{c}_{31}\boldsymbol{x}_1 + \boldsymbol{c}_{32}\boldsymbol{x}_2 + \boldsymbol{c}_{33}\boldsymbol{x}_3 + \boldsymbol{c}_{34}\boldsymbol{x}_4 + \boldsymbol{c}_{35}\boldsymbol{x}_5 \\
\text{subject to} & A_1\boldsymbol{x}_1 + A_2\boldsymbol{x}_2 + A_3\boldsymbol{x}_3 + A_4\boldsymbol{x}_4 + A_5\boldsymbol{x}_5 \le \boldsymbol{b}_0 \\
& B_1\boldsymbol{x}_1 \le \boldsymbol{b}_1 \\
& B_2\boldsymbol{x}_2 \le \boldsymbol{b}_2 \\
& B_3\boldsymbol{x}_3 \le \boldsymbol{b}_3 \\
& B_4\boldsymbol{x}_4 \le \boldsymbol{b}_4 \\
& B_5\boldsymbol{x}_5 \le \boldsymbol{b}_5 \\
& \boldsymbol{x}_j \in \{0,1\}^{n_j}, \ j = 1,2,3,4,5.
\end{array}\right\}$$

As a numerical example with 50 variables, we used the following values of $\boldsymbol{c}_i = (\boldsymbol{c}_{i1}, \boldsymbol{c}_{i2}, \boldsymbol{c}_{i3}, \boldsymbol{c}_{i4}, \boldsymbol{c}_{i5})$, $i = 1,2,3$, A_j and B_j, $j = 1,2,3,4,5$, which were determined at random from the closed intervals.

$$\begin{array}{lrrrrrrrrrrl}
\boldsymbol{c}_1 = (& -27 & -175 & -483 & -475 & -741 & -756 & -141 & -579 & -339 & -776 & \\
& -996 & -209 & -48 & -306 & -66 & -114 & -699 & -122 & -729 & -541 & \\
& -914 & -752 & -713 & -876 & -322 & -462 & -256 & -801 & -196 & -975 & \\
& -69 & -19 & -412 & -207 & -278 & -58 & -157 & -774 & -992 & -203 & \\
& -270 & -597 & -646 & -66 & -636 & -337 & -796 & -851 & -535 & -981 &) \\
\boldsymbol{c}_2 = (& 401 & 875 & 373 & 746 & 844 & 959 & 730 & 538 & 838 & 996 & \\
& 156 & 864 & 412 & 533 & 13 & 131 & 365 & 613 & 370 & 37 & \\
& 381 & 335 & 951 & 860 & 791 & 727 & 132 & 582 & 821 & 595 & \\
& 532 & 243 & 151 & 491 & 221 & 446 & 34 & 472 & 78 & 403 & \\
& 774 & 793 & 334 & 445 & 839 & 198 & 819 & 378 & 385 & 964 &) \\
\boldsymbol{c}_3 = (& -207 & -540 & 719 & -10 & 674 & -508 & -714 & -610 & 583 & 175 & \\
& -371 & -948 & -977 & 345 & 679 & 465 & -925 & -38 & 890 & -266 & \\
& -329 & -99 & 819 & 347 & 867 & -989 & 100 & -258 & -34 & -288 & \\
& -302 & 880 & -335 & 939 & 309 & -899 & -986 & 776 & 882 & -759 & \\
& -180 & -348 & -294 & 375 & -69 & 590 & -452 & -389 & -134 & 358 &)
\end{array}$$

$$A_1 = \begin{pmatrix}
500 & 825 & 351 & 339 & 955 & 30 & 549 & 733 & 543 & 778 \\
149 & 321 & 189 & 873 & 122 & 478 & 598 & 705 & 390 & 590 \\
747 & 368 & 378 & 766 & 740 & 716 & 657 & 35 & 378 & 187 \\
17 & 221 & 944 & 360 & 76 & 375 & 621 & 506 & 323 & 941 \\
451 & 381 & 6 & 754 & 535 & 947 & 437 & 314 & 585 & 737 \\
87 & 466 & 581 & 101 & 594 & 510 & 426 & 638 & 760 & 693 \\
641 & 286 & 932 & 836 & 359 & 316 & 663 & 70 & 691 & 251 \\
276 & 532 & 508 & 865 & 893 & 147 & 965 & 496 & 272 & 370 \\
78 & 140 & 482 & 46 & 580 & 808 & 610 & 314 & 672 & 279 \\
59 & 97 & 881 & 119 & 271 & 820 & 870 & 0 & 142 & 492
\end{pmatrix}$$

$$A_2 = \begin{pmatrix} 326 & 670 & 938 & 948 & 843 & 777 & 435 & 622 & 440 & 772 & 162 & 89 \\ 896 & 754 & 877 & 839 & 185 & 281 & 605 & 244 & 653 & 484 & 201 & 763 \\ 931 & 639 & 826 & 441 & 242 & 885 & 376 & 459 & 582 & 478 & 356 & 433 \\ 915 & 904 & 531 & 376 & 635 & 74 & 802 & 899 & 36 & 25 & 300 & 930 \\ 872 & 734 & 718 & 433 & 643 & 392 & 186 & 648 & 327 & 813 & 141 & 230 \\ 775 & 55 & 643 & 67 & 322 & 226 & 254 & 189 & 92 & 473 & 925 & 681 \\ 964 & 151 & 599 & 758 & 685 & 470 & 585 & 821 & 887 & 294 & 987 & 499 \\ 694 & 686 & 880 & 93 & 652 & 149 & 313 & 814 & 961 & 184 & 197 & 594 \\ 631 & 896 & 123 & 933 & 865 & 973 & 470 & 408 & 135 & 349 & 985 & 338 \\ 503 & 628 & 270 & 309 & 744 & 954 & 763 & 555 & 394 & 118 & 77 & 743 \end{pmatrix}$$

$$A_3 = \begin{pmatrix} 749 & 759 & 311 & 752 & 102 & 624 & 78 & 965 \\ 399 & 644 & 582 & 871 & 827 & 577 & 718 & 895 \\ 356 & 778 & 480 & 335 & 786 & 190 & 800 & 703 \\ 623 & 448 & 813 & 32 & 442 & 493 & 540 & 155 \\ 953 & 674 & 756 & 538 & 320 & 475 & 347 & 694 \\ 899 & 830 & 640 & 825 & 674 & 670 & 277 & 484 \\ 706 & 802 & 14 & 415 & 985 & 807 & 707 & 537 \\ 153 & 544 & 707 & 86 & 212 & 951 & 159 & 416 \\ 695 & 286 & 296 & 107 & 92 & 564 & 324 & 664 \\ 797 & 951 & 636 & 353 & 739 & 901 & 169 & 465 \end{pmatrix}$$

$$A_4 = \begin{pmatrix} 753 & 932 & 875 & 898 & 894 & 348 & 264 & 375 & 890 & 16 & 473 & 563 & 514 \\ 144 & 823 & 427 & 330 & 56 & 329 & 375 & 495 & 305 & 27 & 926 & 971 & 302 \\ 688 & 478 & 901 & 670 & 510 & 266 & 506 & 766 & 45 & 524 & 154 & 529 & 725 \\ 485 & 163 & 36 & 443 & 384 & 860 & 700 & 402 & 44 & 111 & 709 & 695 & 373 \\ 335 & 147 & 652 & 381 & 775 & 403 & 921 & 184 & 372 & 86 & 986 & 19 & 646 \\ 830 & 239 & 143 & 127 & 318 & 115 & 705 & 235 & 528 & 133 & 947 & 877 & 688 \\ 83 & 955 & 443 & 250 & 761 & 697 & 180 & 281 & 558 & 782 & 67 & 476 & 555 \\ 51 & 291 & 537 & 205 & 580 & 20 & 396 & 499 & 31 & 826 & 24 & 901 & 693 \\ 840 & 53 & 692 & 920 & 349 & 582 & 262 & 777 & 983 & 367 & 947 & 290 & 674 \\ 691 & 934 & 151 & 48 & 55 & 414 & 267 & 200 & 569 & 522 & 298 & 121 & 524 \end{pmatrix}$$

$$A_5 = \begin{pmatrix} 343 & 316 & 8 & 905 & 375 & 738 & 62 \\ 673 & 619 & 626 & 230 & 560 & 602 & 515 \\ 313 & 86 & 454 & 902 & 572 & 19 & 311 \\ 14 & 831 & 356 & 756 & 417 & 988 & 710 \\ 320 & 946 & 692 & 369 & 719 & 644 & 98 \\ 808 & 239 & 730 & 925 & 773 & 121 & 331 \\ 593 & 588 & 996 & 939 & 71 & 898 & 926 \\ 49 & 282 & 681 & 707 & 921 & 19 & 145 \\ 807 & 472 & 26 & 990 & 902 & 230 & 127 \\ 223 & 333 & 88 & 954 & 325 & 723 & 223 \end{pmatrix}$$

$$B_1 = \begin{pmatrix} 451 & 655 & 676 & 343 & 765 & 885 & 898 & 644 & 843 & 45 \\ 193 & 542 & 464 & 241 & 423 & 412 & 122 & 623 & 733 & 534 \\ 325 & 125 & 455 & 15 & 950 & 14 & 586 & 972 & 826 & 774 \\ 257 & 751 & 190 & 364 & 92 & 476 & 174 & 372 & 870 & 696 \\ 989 & 100 & 384 & 418 & 944 & 118 & 920 & 307 & 981 & 304 \\ 248 & 333 & 852 & 917 & 541 & 642 & 623 & 721 & 116 & 950 \\ 646 & 697 & 423 & 207 & 650 & 382 & 171 & 857 & 130 & 347 \end{pmatrix}$$

$$B_2 = \begin{pmatrix} 995 & 430 & 658 & 865 & 594 & 785 & 306 & 288 & 717 & 757 & 810 & 170 \\ 923 & 514 & 645 & 662 & 512 & 378 & 668 & 221 & 540 & 416 & 477 & 147 \\ 315 & 571 & 151 & 0 & 289 & 608 & 518 & 403 & 748 & 675 & 617 & 296 \\ 208 & 739 & 507 & 134 & 921 & 692 & 70 & 89 & 398 & 404 & 981 & 779 \\ 359 & 504 & 490 & 548 & 9 & 173 & 168 & 834 & 53 & 723 & 293 & 664 \\ 353 & 256 & 5 & 788 & 632 & 166 & 911 & 504 & 140 & 787 & 283 & 691 \\ 669 & 948 & 501 & 88 & 848 & 732 & 918 & 346 & 103 & 208 & 775 & 891 \\ 204 & 426 & 346 & 123 & 792 & 455 & 958 & 980 & 802 & 735 & 39 & 407 \end{pmatrix}$$

$$B_3 = \begin{pmatrix} 186 & 544 & 868 & 757 & 838 & 837 & 153 & 655 \\ 132 & 267 & 128 & 348 & 977 & 635 & 296 & 94 \\ 796 & 583 & 493 & 598 & 337 & 790 & 422 & 36 \\ 716 & 461 & 309 & 707 & 606 & 386 & 852 & 820 \\ 976 & 355 & 96 & 480 & 279 & 207 & 418 & 374 \end{pmatrix}$$

$$B_4 = \begin{pmatrix} 509 & 298 & 253 & 207 & 19 & 657 & 733 & 413 & 520 & 631 & 892 & 169 & 151 \\ 285 & 584 & 251 & 41 & 343 & 792 & 760 & 267 & 954 & 453 & 244 & 601 & 53 \\ 486 & 251 & 972 & 620 & 159 & 855 & 32 & 596 & 462 & 161 & 743 & 746 & 633 \\ 348 & 391 & 257 & 502 & 263 & 920 & 336 & 807 & 102 & 15 & 610 & 132 & 214 \\ 816 & 592 & 634 & 553 & 691 & 997 & 207 & 809 & 957 & 204 & 239 & 760 & 963 \\ 622 & 622 & 859 & 446 & 260 & 178 & 600 & 554 & 274 & 136 & 695 & 452 & 775 \\ 841 & 579 & 916 & 675 & 335 & 296 & 818 & 859 & 970 & 275 & 908 & 454 & 714 \\ 46 & 806 & 496 & 300 & 409 & 599 & 744 & 880 & 109 & 958 & 959 & 331 & 528 \\ 186 & 401 & 886 & 436 & 82 & 841 & 223 & 329 & 54 & 859 & 692 & 391 & 367 \\ 719 & 815 & 598 & 945 & 596 & 739 & 10 & 622 & 436 & 769 & 774 & 262 & 235 \end{pmatrix}$$

$$B_5 = \begin{pmatrix} 664 & 873 & 639 & 12 & 839 & 144 & 495 \\ 264 & 127 & 612 & 679 & 180 & 268 & 656 \\ 770 & 389 & 770 & 42 & 90 & 546 & 305 \\ 363 & 524 & 501 & 298 & 908 & 39 & 373 \end{pmatrix}$$

Observe that the elements of $\boldsymbol{c}_1$, $\boldsymbol{c}_2$ and $\boldsymbol{c}_3$ were randomly determined from the closed intervals $[-999, 0]$, $[0, 999]$ and $[-999, 999]$, respectively, while the elements of A_j and B_j, $j = 1, 2, 3, 4, 5$, were randomly determined from the closed interval $[0, 999]$.

On the basis of these values, each element $b_j(k)$ of $\boldsymbol{b}_j$, $j = 0, 1, 2, 3, 4, 5$; $k = 1, \ldots, m_j$, was determined by

$$b_j(k) = \begin{cases} \gamma \times \sum_{j=1}^{5} \sum_{j'=1}^{n_j} A^j_{kj'}, \ j = 0 \\ \delta \times \sum_{j'=1}^{n_j} B^j_{kj'}, \qquad j = 1,2,3,4,5 \end{cases}$$

where positive constants γ, δ denote degree of strictness of the coupling constraints and the block constraints respectively, and, $A^j_{kj'}$'s and $B^j_{kj'}$'s denote elements of matrices A_j and B_j respectively.

The numerical experiments were performed on a personal computer (processor: Celeron 466MHz, memory: 128MB, OS: Windows NT 4.0), and a Visual C++ compiler (version 6.0) was used. The parameter values of the genetic algorithm with decomposition procedures were set as, population size = 100, probability of crossover $p_c = 0.8$, probability of mutation $p_m = 0.01$, probability of inversion $p_i = 0.03$, $\varepsilon = 0.01$, $I_{\max} = 1000$ and $I_{\min} = 100$.

Concerning the fuzzy goals of the DM, following Zimmermann [142], for $i = 1,2,3$, after calculating the individual minimum $z_i^{\min}$ together with z_i^{m} for $(\gamma, \delta) = (0.25, 0.25)$, $(0.50, 0.50)$, and $(0.75, 0.75)$, linear membership functions were determined by choosing $z_i^1 = z_i^{\min}$ and $z_i^0 = z_i^{\mathrm{m}}$.

To be more explicit, the values of (z_1^0, z_1^1), (z_2^0, z_2^1) and (z_3^0, z_3^1) were set as $(0, -6848)$, $(5578, 0)$ and $(1692, -4789)$ for $(\gamma, \delta) = (0.25, 0.25)$, $(0, -14947)$, $(11049, 0)$ and $(0, -9974)$ for $(\gamma, \delta) = (0.50, 0.50)$, and $(0, -20469)$, $(17370, 0)$ and $(999, -13040)$ for $(\gamma, \delta) = (0.75, 0.75)$.

Table 8.1 shows approximate optimal solutions obtained by the genetic algorithm with decomposition procedures and exact optimal solutions for three types of (γ, δ), where Best, Average, Worst, and Optimal represent the best value, average value, worst value and optimal solution, respectively.

Table 8.1. Calculation results for fuzzy decision (50 variables)

(γ, δ)	Best	Average	Worst	Optimal	$c_1 x$	$c_2 x$	$c_3 x$	μ_1	μ_2	μ_3
$(0.25, 0.25)$	0.619	0.619	0.619	0.619	−4299	2096	−2322	0.628	0.624	0.619
$(0.50, 0.50)$	0.557	0.554	0.548	0.557	−8437	4898	−5622	0.564	0.557	0.564
$(0.75, 0.75)$	0.582	0.580	0.578	0.582	−11912	7266	−7328	0.582	0.582	0.593

In Table 8.1, observe that the best value obtained through the genetic algorithm with decomposition procedures for each case was equal to the corresponding exact optimal solution obtained by LP_SOLVE [5]. Furthermore, optimal solutions were obtained on 10 times out of 10 trials for $(\gamma, \delta) = (0.25, 0.25)$, while on 4 times out of 10 trials for $(0.50, 0.50)$ and on 1 times out of 10 trials for $(0.75, 0.75)$.

8.4 Interactive fuzzy multiobjective 0-1 programming

8.4.1 Interactive fuzzy programming through genetic algorithms

Having elicited the membership functions $\mu_i(\boldsymbol{c}_i\boldsymbol{x})$, $i = 1, \ldots, k$, from the decision maker (DM) for each of the objective functions $\boldsymbol{c}_i\boldsymbol{x}$, $i = 1, \ldots, k$, the problem (8.1) can be transformed into the following fuzzy multiobjective decision making problem using the general aggregation function $\mu_D(\mu_1(\boldsymbol{c}_1\boldsymbol{x}), \ldots, \mu_k(\boldsymbol{c}_k\boldsymbol{x}))$:

$$\left.\begin{array}{ll} \text{maximize} & \mu_D(\mu_1(\boldsymbol{c}_1\boldsymbol{x}), \ldots, \mu_k(\boldsymbol{c}_k\boldsymbol{x})) \\ \text{subject to} & \boldsymbol{x} \in X \end{array}\right\} \tag{8.10}$$

where the value of the aggregation function $\mu_D(\cdot)$ can be interpreted as the degree of satisfaction of the DM for the whole k fuzzy goals [78].

Probably the most crucial problem in the fuzzy multiobjective decision making problem (8.10) is the identification of an appropriate aggregation function which well represents the human decision makers' fuzzy preferences. If $\mu_D(\cdot)$ can be explicitly identified, then the fuzzy multiobjective decision making problem (8.10) reduces to a standard 0-1 programming problem. However, this seems to happen rarely, and as an alternative approach, interactive fuzzy multiobjective 0-1 programming would be recommended.

To generate a candidate for the satisficing solution which is also Pareto optimal, in interactive fuzzy multiobjective 0-1 programming, the DM is asked to specify the value of the corresponding membership function for each fuzzy goal, called reference membership levels. Observe that the idea of the reference membership levels [78] can be viewed as an obvious extension of the idea of the reference point of Wierzbicki [135, 136]. For the DM's reference membership levels $\bar{\mu}_i$, $i = 1, \ldots, k$, the corresponding Pareto optimal solution, which is nearest to the requirement or better than if the reference membership levels are attainable in the minimax sense, is obtained by solving the following minimax problem in a membership function space [78]:

$$\left.\begin{array}{ll}\text{minimize} & \max_{i=1,\ldots,k}\{\bar{\mu}_i - \mu_i(\boldsymbol{c}_i\boldsymbol{x})\} \\ \text{subject to} & A_1\boldsymbol{x}_1 + \cdots + A_p\boldsymbol{x}_p \le \boldsymbol{b}_0 \\ & B_1\boldsymbol{x}_1 \qquad\qquad\quad \le \boldsymbol{b}_1 \\ & \qquad\quad \ddots \qquad\qquad \vdots \\ & \qquad\qquad\quad B_p\boldsymbol{x}_p \le \boldsymbol{b}_p \\ & \boldsymbol{x}_j \in \{0,1\}^{n_j},\ j=1,\ldots,p. \end{array}\right\} \tag{8.11}$$

It must be noted here that, for generating Pareto optimal solutions by solving the minimax problem, if the uniqueness of the optimal solution is not guaranteed, it is necessary to perform the Pareto optimality test. To circumvent the necessity to perform the Pareto optimality test in the minimax problems, it is reasonable to use the following augmented minimax problem instead of the minimax problem (8.11) [78]:

$$\left.\begin{array}{ll}\text{minimize} & \left\{\max_{i=1,\ldots,k}(\bar{\mu}_i - \mu_i(\boldsymbol{c}_i\boldsymbol{x})) + \rho\sum_{i=1}^{k}(\bar{\mu}_i - \mu_i(\boldsymbol{c}_i\boldsymbol{x}))\right\} \\ \text{subject to} & A_1\boldsymbol{x}_1 + \cdots + A_p\boldsymbol{x}_p \le \boldsymbol{b}_0 \\ & B_1\boldsymbol{x}_1 \qquad\qquad\quad \le \boldsymbol{b}_1 \\ & \qquad\quad \ddots \qquad\qquad \vdots \\ & \qquad\qquad\quad B_p\boldsymbol{x}_p \le \boldsymbol{b}_p \\ & \boldsymbol{x}_j \in \{0,1\}^{n_j},\ j=1,\ldots,p. \end{array}\right\} \tag{8.12}$$

The term augmented is adopted because the term $\rho\sum_{i=1}^{k}(\bar{\mu}_i - \mu_i(\boldsymbol{c}_i\boldsymbol{x}))$ is added to the standard minimax problem, where ρ is a sufficiently small positive number.

Since the constraints of the problem (8.12) preserve the block angular structure, genetic algorithms with decomposition procedures explained in the previous chapter is applicable.

It must be observed here that the fitness of an individual **S** is defined as

$$f(\mathbf{S}) = (1.0 + k\rho) - \max_{i=1,\ldots,k}\left\{(\bar{\mu}_i - \mu_i(\boldsymbol{c}_i\boldsymbol{x})) + \rho\sum_{i=1}^{k}(\bar{\mu}_i - \mu_i(\boldsymbol{c}_i\boldsymbol{x}))\right\} \tag{8.13}$$

where **S** denotes an individual represented by the triple string and $\boldsymbol{x}$ is the phenotype of **S**. For convenience, fitness of an individual **S** is used as fitness of each subindividual $\mathbf{s}^j$, $j = 1,\ldots,p$.

The algorithm of interactive fuzzy multiobjective programming is summarized as follows.

Interactive fuzzy multiobjective programming

Step 0: Calculate the individual minimum and the maximum of each objective function under the given constraints.

Step 1*: Elicit a membership function from the DM for each of the objective functions by taking account of the calculated individual minimum and the maximum of each objective function. Then ask the DM to select the initial reference membership levels $\bar{\mu}_i$, $i = 1, \ldots, k$ (if it is difficult to determine these values, set them to 1).

Step 2: Set an iteration index (generation) $t = 0$ and determine the parameter values for the convergence criterion ε, the minimal search generation $I_{\min}$, the maximal search generation $I_{\max}$, the probability of crossover p_c, the probability of mutation p_m, the probability of inversion p_i.

Step 3: Generate N individuals of the triple string type at random as the initial population.

Step 4: Evaluate each individual (subindividual) on the basis of the phenotype obtained by the proposed decoding algorithm and calculate the mean fitness f_{mean} and the maximal fitness $f_{\max}$ of the population. If $t > I_{\min}$ and $(f_{\max} - f_{\text{mean}})/f_{\max} < \varepsilon$, or, if $t > I_{\max}$, regard the individual with the maximal fitness as the best individual, and stop. Otherwise, set $t = t + 1$ and proceed to Step 5.

Step 5†: Apply the reproduction operator to every subindividual.

Step 6†: Apply the PMX for double strings to the middle and lower part of every subindividual according to the probability of crossover p_c.

Step 7†: Apply the mutation operator of the bit-reverse type to the lower part of every subindividual according to the probability of mutation p_m, and apply the inversion operator for the middle and lower part of every subindividual according to the probability of inversion p_i.

Step 8: Apply the PMX for upper strings according to p_c.

Step 9: Apply the inversion operator for upper strings according to p_i and return to Step 4.

Step 10*: If the DM is satisfied with the current values of the membership functions and/or objective functions given by the current best solution, stop. Otherwise, ask the DM to update the reference membership levels by taking into account the current values of membership functions and/or objective functions, and return to Step 3.

In the algorithm, note that the operations in the steps marked with † can be applied to every subindividual independently. Hence, it becomes possible to reduce the amount of working memory and carry out parallel processing.

8.4.2 Numerical experiments

To illustrate the proposed method, consider the same numerical example discussed in the previous section.

The numerical experiments were performed on a personal computer (processor: Celeron 466MHz, memory: 128MB, OS: Windows NT 4.0) using a Visual C++ compiler (version 6.0). The parameter values of the genetic algorithm with decomposition procedures were set as, population size $= 100$, probability of crossover $p_c = 0.8$, probability of mutation $p_m = 0.01$, probability of inversion $p_i = 0.03$, $\varepsilon = 0.01$, $I_{\max} = 1000$ and $I_{\min} = 100$.

Concerning the fuzzy goals of the decision maker (DM), the values of (z_1^0, z_1^1), (z_2^0, z_2^1) and (z_3^0, z_3^1) were set as $(0, -14947)$, $(11049, 0)$ and $(0, -9974)$ for $(\gamma, \delta) = (0.50, 0.50)$.

As shown in Table 8.2, in the whole interaction process, the values of $(\bar{\mu}_1, \bar{\mu}_2, \bar{\mu}_3)$ were updated from $(1.0, 1.0, 1.0)$ to $(1.0, 0.7, 1.0)$, $(1.0, 0.8, 0.8)$, and $(1.0, 0.8, 0.9)$ sequentially.

For this numerical example, at each interaction with the DM, the corresponding augmented minimax problem was solved through 10 trials of the genetic algorithm with decomposition procedures, as shown in Table 8.2, where # represents the number of the best solution in 10 trials.

The augmented minimax problem was solved for the initial reference membership levels and the DM was supplied with the corresponding Pareto optimal solution and membership values as shown in Interaction 1 of Table 8.2. On the basis of such information, since the DM was not satisfied with the current membership values, the DM updated the reference membership levels to $\bar{\mu}_1 = 1.0$, $\bar{\mu}_2 = 0.7$ and $\bar{\mu}_3 = 1.0$ for improving the satisfaction levels for $\mu_1(\boldsymbol{c}_1\boldsymbol{x})$ and $\mu_3(\boldsymbol{c}_3\boldsymbol{x})$ at the expense of $\mu_2(\boldsymbol{c}_2\boldsymbol{x})$. For the updated reference membership levels, the corresponding augmented minimax problem yielded the Pareto optimal solution and the membership values as shown in Interaction 2 of Table 8.2. Similar procedure continues in this manner until the DM was satisfied with the current values of the membership functions. In this example, at the fourth interaction, the satisficing solution for the DM was derived and the whole interactive process is summarized in Table 8.2.

Table 8.2. Interaction process (10 trials)

Interaction	$(\bar{\mu}_1, \bar{\mu}_2, \bar{\mu}_3)$	c_1x	c_2x	c_3x	$\mu_1(c_1x)$	$\mu_2(c_2x)$	$\mu_3(c_3x)$	#
1st	(1.0, 1.0, 1.0)	−8437	4898	−5622	0.564	0.557	0.564	2
		−8299	4905	−6023	0.555	0.556	0.604	4
		−8289	4840	−5799	0.555	0.562	0.581	2
		−8228	4722	−5472	0.550	0.573	0.549	1
		−8154	4948	−6163	0.546	0.552	0.618	1
2nd	(1.0, 0.7, 1.0)	−10063	6971	−6690	0.673	0.369	0.671	2
		−10065	6995	−6675	0.673	0.367	0.669	1
		−9947	6971	−6830	0.665	0.369	0.685	2
		−9920	7038	−6815	0.664	0.363	0.683	2
		−9907	7021	−6846	0.663	0.365	0.686	1
		−9878	7005	−6659	0.661	0.366	0.668	1
		−9744	6990	−6496	0.652	0.367	0.651	1
3rd	(1.0, 0.8, 0.8)	−10161	5576	−4902	0.680	0.495	0.491	2
		−10537	5795	−4876	0.705	0.476	0.489	1
		−10081	5732	−5242	0.674	0.481	0.526	1
		−10064	5729	−5022	0.673	0.481	0.504	1
		−10210	5776	−4708	0.683	0.477	0.472	1
		−10033	5744	−4708	0.671	0.480	0.472	1
		−10019	5746	−4725	0.670	0.480	0.474	1
		−10006	5554	−4818	0.669	0.497	0.483	1
4th	(1.0, 0.8, 0.9)	−10047	5942	−5649	0.672	0.462	0.566	1
		−9902	5985	−5789	0.662	0.458	0.580	1
		−9830	5992	−5684	0.658	0.458	0.570	1
		−9813	5909	−5538	0.657	0.465	0.555	1
		−9791	6036	−5629	0.655	0.454	0.564	1
		−9771	6065	−5538	0.654	0.451	0.555	2
		−9704	5883	−5918	0.650	0.468	0.593	2
		−9717	6151	−5419	0.650	0.443	0.543	1

Observe that the average processing time in 10 trials were 93.47 seconds and the standard deviation was 0.22.

9. Large Scale Interactive Multiobjective 0-1 Programming with Fuzzy Numbers

In this chapter, as the 0-1 version of Chapter 6, large scale multiobjective multidimensional 0-1 knapsack problems with block angular structures involving fuzzy numbers are formulated. Along the same line as in Chapter 6, interactive decision making methods using the genetic algorithm with decomposition procedures, both without and with the fuzzy goals of the decision maker, for deriving a satisficing solution efficiently from an extended Pareto optimal solution set are presented. Several numerical experiments are also given.

9.1 Introduction

In the previous chapter, we focused mainly on large scale multiobjective 0-1 knapsack problems with block angular structures and presented fuzzy multiobjective 0-1 programming, and interactive fuzzy multiobjective 0-1 programming to derive the satisficing solution of the decision maker (DM) efficiently from a Pareto optimal solution set.

However, as discussed in Chapter 6, when formulating large scale multiobjective multidimensional 0-1 knapsack problems with block angular structures which closely describe and represent the real-world decision situations, various factors of the real-world systems should be reflected in the description of the objective functions and the constraints. Naturally, these objective functions and constraints involve many parameters whose possible values may be assigned by the experts. In the conventional approaches, such parameters are required to fix at some values in an experimental and/or subjective manner through the experts' understanding of the nature of the parameters in the problem-formulation process.

It must be observed here that, in most real-world situations, the possible values of these parameters are often only imprecisely or ambiguously known to the experts. With this observation, it would be certainly more appropriate to interpret the experts' understanding of the parameters as fuzzy numerical data which can be represented by means of fuzzy sets of the real line known as fuzzy numbers. The resulting multiobjective 0-1 knapsack problems with block angular structures involving fuzzy parameters would be viewed as more realistic versions than the conventional ones.

For dealing with large scale multiobjective 0-1 knapsack problems with block angular structures involving fuzzy parameters characterized by fuzzy numbers, Kato and Sakawa [49, 50, 51] recently presented two types of interactive decision making methods, using genetic algorithms with decomposition procedures, which are applicable and promising for handling and tackling not only the experts's fuzzy understanding of the nature of parameters in the problem-formulation process but also the fuzzy goals of the DM. These methods can be viewed as a natural generalization of the previous results for block angular multiobjective 0-1 knapsack problems without fuzzy parameters by Kato and Sakawa [46, 47, 49]. In this chapter, these two types of interactive decision making methods for large scale multiobjective 0-1 knapsack problems with block angular structures involving fuzzy numbers to derive the satisficing solution for the DM efficiently from the extended Pareto optimal solution set are presented in detail together with several numerical experiments.

9.2 Interactive multiobjective 0-1 programming with fuzzy numbers

9.2.1 Problem formulation and solution concepts

As discussed in the previous chapter, consider the large scale multiobjective multidimensional 0-1 knapsack problem of the following block angular form [46, 47]:

$$\left.\begin{array}{lrl}
\text{minimize} & \boldsymbol{c}_1\boldsymbol{x} = & \boldsymbol{c}_{11}\boldsymbol{x}_1 + \cdots + \boldsymbol{c}_{1p}\boldsymbol{x}_p \\
\vdots & & \vdots \\
\text{minimize} & \boldsymbol{c}_k\boldsymbol{x} = & \boldsymbol{c}_{k1}\boldsymbol{x}_1 + \cdots + \boldsymbol{c}_{kp}\boldsymbol{x}_p \\
\text{subject to} & & A_1\boldsymbol{x}_1 + \cdots + A_p\boldsymbol{x}_p \le \boldsymbol{b}_0 \\
 & & B_1\boldsymbol{x}_1 \qquad\qquad\quad \le \boldsymbol{b}_1 \\
 & & \ddots \qquad\qquad \vdots \\
 & & B_p\boldsymbol{x}_p \le \boldsymbol{b}_p \\
 & & \boldsymbol{x}_j \in \{0,1\}^{n_j},\ j = 1,\ldots,p
\end{array}\right\} \quad (9.1)$$

where $\boldsymbol{c}_{ij}$, $i = 1,\ldots,k$, $j = 1,\ldots,p$, are n_j dimensional cost factor row vectors, $\boldsymbol{x}_j$, $j = 1,\ldots,p$, are n_j dimensional column vectors of 0-1 decision variables, $A_1\boldsymbol{x}_1 + \cdots + A_p\boldsymbol{x}_p \le \boldsymbol{b}_0$ denotes m_0 dimensional coupling constraints, and A_j, $j = 1,\ldots,p$, are $m_0 \times n_j$ coefficient matrices. The inequalities $B_j\boldsymbol{x}_j \le \boldsymbol{b}_j$, $j = 1,\ldots,p$, are m_j dimensional block constraints with respect to $\boldsymbol{x}_j$ and B_j, $j = 1,\ldots,p$, are $m_j \times n_j$ coefficient matrices. Here as in the previous chapter, it is assumed that all elements of A_j, B_j and $\boldsymbol{b}_j$ are nonnegative.

For simplicity in notations, define the following vectors and matrices:

$$\boldsymbol{c}_i = (\boldsymbol{c}_{i1},\ldots,\boldsymbol{c}_{ip}),\ i = 1,\ldots,k,$$

$$\begin{aligned}
\boldsymbol{c} &= (\boldsymbol{c}_1,\ldots,\boldsymbol{c}_k),\\
\boldsymbol{x} &= (\boldsymbol{x}_1^T,\ldots,\boldsymbol{x}_p^T),\\
A = (A_1,\ldots,A_p),\ B &= \begin{bmatrix} B_1 & & O\\ & \ddots & \\ O & & B_p\end{bmatrix},\\
\boldsymbol{b} &= (\boldsymbol{b}_0^T,\boldsymbol{b}_1^T,\ldots,\boldsymbol{b}_p^T)^T.
\end{aligned}$$

However, it would certainly be more appropriate to consider that the possible values of the parameters in the description of the objective functions and the constraints of the large scale multiobjective multidimensional 0-1 knapsack problem, although in the conventional approaches they are fixed at some values, usually involve the ambiguity of the experts' understanding of the real system in the problem-formulation process. For this reason, in this chapter, we consider the following large scale multiobjective multidimensional 0-1 knapsack problem with the block angular structure involving fuzzy parameters:

$$\left.\begin{array}{lrcl}
\text{minimize} & \tilde{\boldsymbol{c}}_1\boldsymbol{x} = \tilde{\boldsymbol{c}}_{11}\boldsymbol{x}_1 & + \cdots + & \tilde{\boldsymbol{c}}_{1p}\boldsymbol{x}_p \\
\vdots & \vdots & & \\
\text{minimize} & \tilde{\boldsymbol{c}}_k\boldsymbol{x} = \tilde{\boldsymbol{c}}_{k1}\boldsymbol{x}_1 & + \cdots + & \tilde{\boldsymbol{c}}_{kp}\boldsymbol{x}_p \\
\text{subject to} & \tilde{A}_1\boldsymbol{x}_1 & + \cdots + & \tilde{A}_p\boldsymbol{x}_p \le \tilde{\boldsymbol{b}}_0 \\
 & \tilde{B}_1\boldsymbol{x}_1 & & \le \tilde{\boldsymbol{b}}_1 \\
 & & \ddots & \vdots \\
 & & & \tilde{B}_p\boldsymbol{x}_p \le \tilde{\boldsymbol{b}}_p \\
 & \boldsymbol{x}_j \in \{0,1\}^{n_j}, & j = 1,\ldots,p &
\end{array}\right\} \tag{9.2}$$

where $\tilde{A}_j$ and $\tilde{B}_j$ or $\tilde{\boldsymbol{b}}_j$ and $\tilde{\boldsymbol{c}}_{ij}$ respectively represent matrices or vectors whose elements are fuzzy parameters. These fuzzy parameters are assumed to be characterized as fuzzy numbers defined as Definition 2.2.1. It is assumed here that all of the fuzzy numbers in $\tilde{A}_j$, $\tilde{B}_j$, $\tilde{\boldsymbol{b}}_j$ and $\tilde{\boldsymbol{c}}_{ij}$ are nonnegative. Then the problem (9.2) can be viewed as a multiobjective multidimensional 0-1 knapsack problem with a block angular structure involves fuzzy numbers.

Observing that the problem (9.2) involves fuzzy numbers, it is evident that the concept of Pareto optimality defined for multiobjective 0-1 programming problems cannot be applied directly. Thus, it seems essential to extend the concept of the usual Pareto optimality in some sense. For that purpose, we introduce the α-level set of fuzzy numbers.

Definition 9.2.1 (α-level set).

The α-level set of $\tilde{A}$, $\tilde{B}$, $\tilde{\boldsymbol{b}}$ and $\tilde{\boldsymbol{c}}$ is defined as the ordinary set $(\tilde{A}, \tilde{B}, \tilde{\boldsymbol{b}}, \tilde{\boldsymbol{c}})_\alpha$ for which the degree of their membership functions exceeds the level α:

$$
\begin{aligned}
&(\tilde{A}, \tilde{B}, \tilde{\boldsymbol{b}}, \tilde{\boldsymbol{c}})_\alpha \\
&= \{(A, B, \boldsymbol{b}, \boldsymbol{c}) \mid \mu_{\tilde{a}^s_{ij}}(a^s_{ij}) \geq \alpha, \mu_{\tilde{b}^s_{hj}}(b^s_{hj}) \geq \alpha, \mu_{\tilde{b}'^{s'}_{h'}}(b'^{s'}_{h'}) \geq \alpha, \mu_{\tilde{c}^{ts}_j}(c^{ts}_j) \geq \alpha, \\
&\qquad i = 1, \ldots, m_0, j = 1, \ldots, n_s, s = 1, \ldots, p, h = 1, \ldots, m_s, \\
&\qquad h' = 1, \ldots, m_{s'}, s' = 0, \ldots, p, t = 1, \ldots, k\}
\end{aligned}
$$

Now, as discussed in Chapter 6, suppose that the decision maker (DM) considers that the degree of all of the membership functions of the fuzzy numbers involved in the large scale multiobjective multidimensional 0-1 knapsack problem with the block angular structure involving fuzzy numbers should be greater than or equal to a certain value of α. Then, for such a degree α, the large scale multiobjective multidimensional 0-1 knapsack problem with the block angular structure involving fuzzy numbers can be interpreted as the following nonfuzzy α-multiobjective 0-1 programming problem:

$$
\left.
\begin{array}{lrl}
\text{minimize} & \boldsymbol{c}_1\boldsymbol{x} = & \boldsymbol{c}_{11}\boldsymbol{x}_1 + \cdots + \boldsymbol{c}_{1p}\boldsymbol{x}_p \\
\vdots & & \vdots \\
\text{minimize} & \boldsymbol{c}_k\boldsymbol{x} = & \boldsymbol{c}_{k1}\boldsymbol{x}_1 + \cdots + \boldsymbol{c}_{kp}\boldsymbol{x}_p \\
\text{subject to} & & A_1\boldsymbol{x}_1 + \cdots + A_p\boldsymbol{x}_p \leq \boldsymbol{b}_0 \\
 & & B_1\boldsymbol{x}_1 \qquad\qquad\quad \leq \boldsymbol{b}_1 \\
 & & \qquad \ddots \qquad\qquad \vdots \\
 & & \qquad\qquad B_p\boldsymbol{x}_p \leq \boldsymbol{b}_p \\
 & & \boldsymbol{x}_j \in \{0,1\}^{n_j},\ j = 1, \ldots, p \\
 & & (A, B, \boldsymbol{b}, \boldsymbol{c}) \in (\tilde{A}, \tilde{B}, \tilde{\boldsymbol{b}}, \tilde{\boldsymbol{c}})_\alpha
\end{array}
\right\} \tag{9.3}
$$

where $X(A, B, \boldsymbol{b})$ denotes the feasible region of the above α-multiobjective 0-1 programming problem. It should be emphasized that, in this formulation, the parameters $(A, B, \boldsymbol{b}, \boldsymbol{c})$ are treated as decision variables rather than constants.

On the basis of the α-level sets of the fuzzy numbers, the concept of an α-Pareto optimal solution can be introduced to the α-multiobjective 0-1 programming problem as a natural extension of the Pareto optimality concept for the ordinary multiobjective 0-1 programming problem.

Definition 9.2.2 (α-Pareto optimal solution).
$\boldsymbol{x}^* \in X(A^*, B^*, \boldsymbol{b}^*)$ is an α-Pareto optimal solution to the α-multiobjective 0-1 programming problem, if and only if there does not exist another $\boldsymbol{x} \in X(A, B, \boldsymbol{b})$, $(A, B, \boldsymbol{b}, \boldsymbol{c}) \in (\tilde{A}, \tilde{B}, \tilde{\boldsymbol{b}}, \tilde{\boldsymbol{c}})_\alpha$ such that $\boldsymbol{c}_i\boldsymbol{x} \le \boldsymbol{c}_i^*\boldsymbol{x}^*$, $i = 1, \dots, k$, and $\boldsymbol{c}_j\boldsymbol{x} < \boldsymbol{c}_j^*\boldsymbol{x}^*$ for at least one j, $1 \le j \le k$. The corresponding values of parameters $(A^*, B^*, \boldsymbol{b}^*, \boldsymbol{c}^*)$ are called α-level optimal parameters.

As can be immediately understood from Definition 9.2.2, in general, the α-Pareto optimal solution to the α-multiobjective 0-1 programming problem consists of an infinite number of points. Therefore some subjective judgments should be added to the quantitative analyses by the DM. Namely, the DM must select a compromise or satisficing solution from the α-Pareto optimal solution set, based on DM's subjective value judgments.

9.2.2 Interactive programming through genetic algorithms

To generate a candidate for the satisficing solution which is also α-Pareto optimal, in our interactive decision making method, the DM is asked to specify the degree α of the α-level set and the reference levels of achievement of the objective functions. Given a degree α and reference levels $\bar{z}_i$, $i = 1, \dots, k$, the corresponding α-Pareto optimal solution, which is nearest to the requirement or better than if the reference levels are attainable in the minimax sense, is obtained by solving the following minimax problem in an objective function space [78]:

$$\left.\begin{array}{ll}
\text{minimize} & \displaystyle\max_{i=1,\dots,k} \{\boldsymbol{c}_i\boldsymbol{x} - \bar{z}_i\} \\
\text{subject to} & \begin{array}{rcl} A_1\boldsymbol{x}_1 + \cdots + A_p\boldsymbol{x}_p & \le & \boldsymbol{b}_0 \\ B_1\boldsymbol{x}_1 \qquad\qquad & \le & \boldsymbol{b}_1 \\ \ddots \qquad & & \vdots \\ B_p\boldsymbol{x}_p & \le & \boldsymbol{b}_p \end{array} \\
& \boldsymbol{x}_j \in \{0,1\}^{n_j}, \; j = 1, \dots, p \\
& (A, B, \boldsymbol{b}, \boldsymbol{c}) \in (\tilde{A}, \tilde{B}, \tilde{\boldsymbol{b}}, \tilde{\boldsymbol{c}})_\alpha
\end{array}\right\} \tag{9.4}$$

It should be noted here that, for generating α-Pareto optimal solutions by solving the minimax problem (9.4), the uniqueness of the optimal solution must be verified. For circumventing the α-Pareto optimality tests as in the minimax problems, it is recommended to use the following augmented minimax problem rather than the minimax problem (9.4) [78].

$$\left.\begin{array}{ll}\text{minimize} & \left\{\max\limits_{i=1,\dots,k}(\boldsymbol{c}_i\boldsymbol{x}-\bar{z}_i)+\rho\sum\limits_{i=1}^{k}(\boldsymbol{c}_i\boldsymbol{x}-\bar{z}_i)\right\} \\ \text{subject to} & \begin{array}{rcl} A_1\boldsymbol{x}_1 + \cdots + A_p\boldsymbol{x}_p & \le & \boldsymbol{b}_0 \\ B_1\boldsymbol{x}_1 \qquad\qquad & \le & \boldsymbol{b}_1 \\ \ddots \qquad & & \vdots \\ B_p\boldsymbol{x}_p & \le & \boldsymbol{b}_p \end{array} \\ & \boldsymbol{x}_j \in \{0,1\}^{n_j},\ j=1,\dots,p \\ & (A,B,\boldsymbol{b},\boldsymbol{c}) \in (\tilde{A},\tilde{B},\tilde{\boldsymbol{b}},\tilde{\boldsymbol{c}})_\alpha \end{array}\right\} \quad (9.5)$$

where ρ is a sufficiently small positive number.

In this formulation, however, constraints are nonlinear because the parameters A, B, $\boldsymbol{b}$ and $\boldsymbol{c}$ are treated as decision variables. To deal with such nonlinearities, we introduce the following set-valued functions $S_i(\cdot)$, $T_j(\cdot,\cdot)$ and $U(\cdot,\cdot)$:

$$S_i(\boldsymbol{c}_i) = \left\{(\boldsymbol{x},v) \;\middle|\; \boldsymbol{c}_i\boldsymbol{x}-\bar{z}_i+\rho\sum_{i=1}^{k}(\boldsymbol{c}_i\boldsymbol{x}-\bar{z}_i)\le v,\ x_{ij}\in\{0,1\}\right\},$$
$$T_j(B_j,\boldsymbol{b}_j) = \{\boldsymbol{x}_j \mid B_j\boldsymbol{x}_j \le \boldsymbol{b}_j,\ x_{ij}\in\{0,1\}\},$$
$$U(A,\boldsymbol{b}_0) = \{\boldsymbol{x} \mid A\boldsymbol{x} \le \boldsymbol{b}_0,\ x_{ij}\in\{0,1\}\}.$$

Then it can be easily verified that the following relations hold for $T_j(\cdot,\cdot)$ and $U(\cdot,\cdot)$ when $\boldsymbol{x} \ge \boldsymbol{0}$.

Proposition 9.2.1.
(1) *If* $\boldsymbol{c}_i^1 \le \boldsymbol{c}_i^2$, *then* $S_i(\boldsymbol{c}_i^1) \supseteq S_i(\boldsymbol{c}_i^2)$,
(2) *If* $B_j^1 \le B_j^2$, *then* $T_j(B_j^1,\boldsymbol{b}_j) \supseteq T_j(B_j^2,\boldsymbol{b}_j)$,
(3) *If* $\boldsymbol{b}_j^1 \le \boldsymbol{b}_j^2$, *then* $T_j(B_j,\boldsymbol{b}_j^1) \subseteq T_j(B_j,\boldsymbol{b}_j^2)$,
(4) *If* $A^1 \le A^2$, *then* $U(A^1,\boldsymbol{b}_0) \supseteq U(A^2,\boldsymbol{b}_0)$,
(5) *If* $\boldsymbol{b}_0^1 \le \boldsymbol{b}_0^2$, *then* $U(A,\boldsymbol{b}^1) \subseteq U(A,\boldsymbol{b}^2)$.

Now from the properties of the α-level set for the vectors and/or matrices of fuzzy numbers, the feasible regions for $\boldsymbol{c}_i$, A, B_j and $\boldsymbol{b}_j$ can be denoted respectively by the closed intervals $[\boldsymbol{c}_{i\alpha}^L, \boldsymbol{c}_{i\alpha}^R]$, $[A_\alpha^L, A_\alpha^R]$, $[B_{j\alpha}^L, B_{j\alpha}^R]$

and $[b^L_{j\alpha}, b^R_{j\alpha}]$, where y^L_α or y^R_α represents the left or right extreme point of the α-level set $\tilde{y}_\alpha$.

Therefore, through the use of Proposition 5.1, we can obtain the optimal solution of the problem (9.5) by solving the following problem:

$$\left.\begin{array}{ll}\text{minimize} & \left\{\max\limits_{i=1,\ldots,k}(\boldsymbol{c}^L_{i\alpha}\boldsymbol{x}-\bar{z}_i)+\rho\sum\limits_{i=1}^{k}(\boldsymbol{c}^L_{i\alpha}\boldsymbol{x}-\bar{z}_i)\right\} \\ \text{subject to} & A^L_{1\alpha}\boldsymbol{x}_1 + \cdots + A^L_{p\alpha}\boldsymbol{x}_p \le \boldsymbol{b}^R_{0\alpha} \\ & B^L_{1\alpha}\boldsymbol{x}_1 \qquad\qquad\quad \le \boldsymbol{b}^R_{1\alpha} \\ & \qquad\ddots \qquad\qquad\qquad \vdots \\ & \qquad\qquad B^L_{p\alpha}\boldsymbol{x}_p \le \boldsymbol{b}^R_{p\alpha} \\ & \boldsymbol{x}_j \in \{0,1\}^{n_j},\ j=1,\ldots,p \end{array}\right\} \tag{9.6}$$

Fortunately, the constraints of the problem (9.6) preserve the block angular structure, genetic algorithms with decomposition procedures explained in Chapter 7 is applicable for solving the problem (9.6).

Here, the fitness of an individual **S** is defined by

$$f(\mathbf{S}) = \min_{i=1,\ldots,k}\{\bar{z}_i - \boldsymbol{c}^L_{i\alpha}\boldsymbol{x}\} + \rho\sum_{i=1}^{k}(\bar{z}_i - \boldsymbol{c}^L_{i\alpha}\boldsymbol{x}) - (1+k\rho)(\bar{z}_{\min} - z_{\max}) \tag{9.7}$$

where $\boldsymbol{x}$ is the solution (phenotype) generated from the individual **S** (genotype) by the decoding algorithm, and $\bar{z}_{\min}$ and $z_{\max}$ are respectively defined as

$$\bar{z}_{\min} = \min_{i=1,\ldots,k}\{\bar{z}_i\} \tag{9.8}$$

$$z_{\max} = \max_{i=1,\ldots,k}\max_{\boldsymbol{x}\in X}\{\boldsymbol{c}^L_{i\alpha}\boldsymbol{x} \mid \alpha = 0\} \tag{9.9}$$

Following the preceding discussions, we can now construct the interactive algorithm in order to derive the satisficing solution for the DM from the α-Pareto optimal solution set.

Interactive multiobjective 0-1 programming with fuzzy numbers

Step 0: Calculate the individual minimum and maximum of each objective function under the given constraints for $\alpha = 0$ and $\alpha = 1$ respectively.

Step 1*: Ask the DM to select the initial value of α $(0 \le \alpha \le 1)$ and the initial reference levels $\bar{z}_i$, $i = 1, \ldots, k$.

Step 2: Set the iteration index (generation) $t = 0$ and determine the parameter values for the convergence criterion ε, the minimal search generation $I_{\min}$, the maximal search generation $I_{\max}$, the probability of crossover p_c, the probability of mutation p_m, the probability of inversion p_i.

Step 3: Generate N individuals of the triple string type at random as the initial population.

Step 4: Evaluate each individual (subindividual) on the basis of the phenotype obtained by the proposed decoding algorithm, and calculate the mean fitness f_{mean} and the maximal fitness $f_{\max}$ of the population. If $t > I_{\min}$ and $(f_{\max} - f_{\text{mean}})/f_{\max} < \varepsilon$, or, if $t > I_{\max}$, regard the individual with the maximal fitness as the best individual, and stop. Otherwise, set $t = t + 1$ and proceed to Step 5.

Step 5†: Apply the reproduction operator to every subindividual.

Step 6†: Apply the PMX for double strings to the middle and lower part of every subindividual according to the probability of crossover p_c.

Step 7†: Apply the mutation operator of the bit-reverse type to the lower part of every subindividual according to the probability of mutation p_m, apply the inversion operator for the middle and lower part of every subindividual according to the probability of inversion p_i.

Step 8: Apply the PMX for upper strings according to p_c.

Step 9: Apply the inversion operator for upper strings according to p_i and return to Step 4.

Step 10*: If the DM is satisfied with the current values of the objective functions given by the current best solution, stop. Otherwise, ask the DM to update the reference levels and α by taking into account the current values of the objective functions and return to Step 3.

Observe that the operations in the steps marked with † can be applied to every subindividual independently. Accordingly, it is theoretically possible to reduce the amount of working memory and carry out parallel processing.

9.2.3 Numerical experiments

To illustrate the proposed method, consider the following multiobjective 0-1 programming problem with the block angular structure involving fuzzy numbers (3 objectives, 4 blocks, 30 (=4+9+10+7) variables and 3 coupling constraints):

$$
\left.\begin{array}{llllll}
\text{minimize} & \tilde{\boldsymbol{c}}_{11}\boldsymbol{x}_1 & + \tilde{\boldsymbol{c}}_{12}\boldsymbol{x}_2 & + \tilde{\boldsymbol{c}}_{13}\boldsymbol{x}_3 & + \tilde{\boldsymbol{c}}_{14}\boldsymbol{x}_4 & \\
\text{minimize} & \tilde{\boldsymbol{c}}_{21}\boldsymbol{x}_1 & + \tilde{\boldsymbol{c}}_{22}\boldsymbol{x}_2 & + \tilde{\boldsymbol{c}}_{23}\boldsymbol{x}_3 & + \tilde{\boldsymbol{c}}_{24}\boldsymbol{x}_4 & \\
\text{minimize} & \tilde{\boldsymbol{c}}_{31}\boldsymbol{x}_1 & + \tilde{\boldsymbol{c}}_{32}\boldsymbol{x}_2 & + \tilde{\boldsymbol{c}}_{33}\boldsymbol{x}_3 & + \tilde{\boldsymbol{c}}_{34}\boldsymbol{x}_4 & \\
\text{subject to} & \tilde{A}_1\boldsymbol{x}_1 & + \tilde{A}_2\boldsymbol{x}_2 & + \tilde{A}_3\boldsymbol{x}_3 & + \tilde{A}_4\boldsymbol{x}_4 & \leq \tilde{\boldsymbol{b}}_0 \\
 & \tilde{B}_1\boldsymbol{x}_1 & & & & \leq \tilde{\boldsymbol{b}}_1 \\
 & & \tilde{B}_2\boldsymbol{x}_2 & & & \leq \\
\mathit{tildevcb}_2 & & & & & \\
 & & & \tilde{B}_3\boldsymbol{x}_3 & & \leq \tilde{\boldsymbol{b}}_3 \\
 & & & & \tilde{B}_4\boldsymbol{x}_4 & \leq \tilde{\boldsymbol{b}}_4 \\
 & \multicolumn{5}{l}{\boldsymbol{x}_j \in \{0,1\}^{n_j}, \ j = 1,2,3,4.}
\end{array}\right\} \qquad (9.10)
$$

As a numerical example with 30 variables, we used the following values of $\tilde{\boldsymbol{c}}_i = (\tilde{\boldsymbol{c}}_{i1}, \tilde{\boldsymbol{c}}_{i2}, \tilde{\boldsymbol{c}}_{i3}, \tilde{\boldsymbol{c}}_{i4})$, $i = 1,2,3$, $\tilde{A}_j$ and $\tilde{B}_j$, $j = 1,2,3,4$. Observe that the elements of $\tilde{\boldsymbol{c}}_1$, $\tilde{\boldsymbol{c}}_2$ and $\tilde{\boldsymbol{c}}_3$ were randomly determined from the closed intervals $[-999, 0]$, $[0, 999]$ and $[-999, 999]$, respectively, while the elements of $\tilde{A}_j$ and $\tilde{B}_j$, $j = 1,2,3,4$, were randomly determined from the closed interval $[0, 999]$, where the numbers with tildes are triangular fuzzy numbers.

$$
\begin{array}{rl}
\tilde{\boldsymbol{c}}_1 = (& -615 \ \ -172 \ \ -\widetilde{554} \ \ -292 \ \ -872 \ \ -835 \ \ -844 \ \ -895 \ \ -594 \ \ -540 \\
 & -168 \ \ -654 \ \ -\widetilde{690} \ \ -263 \ \ -106 \ \ -\widetilde{814} \ \ -191 \ \ -423 \ \ -351 \ \ -839 \\
 & -137 \ \ -262 \ \ -177 \ \ -479 \ \ -380 \ \ -504 \ \ -502 \ \ -351 \ \ -525 \ \ -120 \ ) \\
\tilde{\boldsymbol{c}}_2 = (& 519 \ \ 607 \ \ 732 \ \ 556 \ \ 344 \ \ \widetilde{801} \ \ \widetilde{590} \ \ \widetilde{266} \ \ 670 \ \ 552 \\
 & 788 \ \ 887 \ \ 889 \ \ 68 \ \ 800 \ \ 907 \ \ 644 \ \ \widetilde{165} \ \ 301 \ \ 166 \\
 & 285 \ \ \widetilde{841} \ \ 536 \ \ 36 \ \ 207 \ \ 21 \ \ 358 \ \ \widetilde{621} \ \ 520 \ \ 546 \ ) \\
\tilde{\boldsymbol{c}}_3 = (& -153 \ \ 33 \ \ -378 \ \ -20 \ \ 915 \ \ 729 \ \ -982 \ \ 112 \ \ -\widetilde{799} \ \ 738 \\
 & 739 \ \ 362 \ \ 183 \ \ 115 \ \ \widetilde{788} \ \ -790 \ \ \widetilde{402} \ \ -\widetilde{182} \ \ -331 \ \ 964 \\
 & 184 \ \ 54 \ \ -10 \ \ -\widetilde{958} \ \ -955 \ \ -39 \ \ 146 \ \ 197 \ \ -26 \ \ -\widetilde{919} \ )
\end{array}
$$

$$
\tilde{A}_1 = \begin{pmatrix} 865 & \widetilde{148} & 171 & 68 \\ 651 & 736 & 102 & 160 \\ 93 & 121 & 24 & \widetilde{762} \end{pmatrix}, \ \tilde{A}_2 = \begin{pmatrix} 956 & 27 & 646 & 108 & 427 & 309 & 18 & \widetilde{885} & 757 \\ 509 & 165 & \widetilde{762} & 880 & 499 & 875 & 734 & 235 & 51 \\ 605 & 875 & 504 & 678 & 989 & \widetilde{604} & 496 & 589 & 895 \end{pmatrix},
$$

$$
\tilde{A}_3 = \begin{pmatrix} 44 & 882 & 108 & 520 & \widetilde{578} & 9 & 387 & 477 & 192 & 507 \\ 775 & 354 & 697 & 912 & 670 & \widetilde{705} & 426 & 20 & 212 & 947 \\ 502 & 194 & 644 & 127 & 264 & 336 & 703 & 38 & 953 & 754 \end{pmatrix},
$$

$$
\tilde{A}_4 = \begin{pmatrix} 874 & 634 & 243 & 635 & \widetilde{850} & 237 & 720 \\ 339 & 50 & 485 & 897 & 242 & 527 & 494 \\ 855 & \widetilde{345} & 123 & 215 & 115 & 363 & 204 \end{pmatrix}
$$

$$\tilde{B}_1 = \begin{pmatrix} 513 & \widetilde{175} & 308 & 534 \\ 947 & 171 & \widetilde{702} & 226 \\ \widetilde{494} & 124 & 83 & 389 \\ 277 & 368 & 983 & \widetilde{535} \end{pmatrix}, \quad \tilde{B}_2 = \begin{pmatrix} 765 & 644 & 767 & \widetilde{780} & 822 & 151 & 625 & 314 & 346 \\ 917 & 519 & 401 & 606 & 785 & 931 & \widetilde{869} & 866 & 674 \\ \widetilde{758} & 581 & 389 & 355 & 200 & 826 & 415 & \widetilde{463} & 979 \\ 126 & 212 & 958 & 737 & 409 & \widetilde{780} & 757 & 956 & 28 \\ 318 & 756 & \widetilde{242} & 589 & 43 & 956 & 319 & 59 & 441 \\ 915 & \widetilde{572} & 118 & 569 & 252 & 495 & 236 & 476 & 406 \\ 872 & 426 & 358 & 381 & 43 & 160 & 522 & \widetilde{696} & 97 \end{pmatrix}$$

$$\tilde{B}_3 = \begin{pmatrix} 400 & 773 & 244 & 342 & 229 & 297 & 304 & 887 & 36 & \widetilde{651} \\ \widetilde{398} & 676 & 732 & 937 & 233 & 838 & 967 & 778 & 431 & 674 \\ 809 & 158 & \widetilde{279} & 135 & 864 & 750 & 207 & 139 & 294 & 802 \\ 218 & \widetilde{563} & 715 & 197 & 989 & 250 & 430 & 755 & 860 & 894 \\ 978 & 395 & 432 & 127 & \widetilde{457} & 237 & 986 & 652 & 604 & 241 \end{pmatrix}$$

$$\tilde{B}_4 = \begin{pmatrix} 454 & 789 & 78 & \widetilde{476} & 152 & 245 & 944 \\ 614 & 988 & 477 & 799 & \widetilde{744} & 380 & 479 \\ 526 & 98 & \widetilde{594} & 347 & 143 & 779 & 710 \\ 446 & 704 & 95 & 962 & 551 & 740 & \widetilde{579} \\ \widetilde{615} & 172 & 554 & 292 & 282 & 684 & 292 \\ 872 & \widetilde{835} & 844 & 895 & 594 & 540 & 168 \\ 263 & 106 & 814 & 191 & 423 & \widetilde{351} & 839 \\ 479 & 380 & \widetilde{504} & 502 & 351 & 525 & 120 \end{pmatrix}$$

Each element $\tilde{b}_j(k)$ of $\tilde{\boldsymbol{b}}_j$ was determined by

$$\begin{cases} \tilde{b}_0(k) = 0.5 \times \displaystyle\sum_{j=1}^{4} \sum_{j'=1}^{n_j} \tilde{A}^j_{kj'} \\ \tilde{b}_j(k) = 0.5 \times \displaystyle\sum_{j'=1}^{n_j} \tilde{B}^j_{kj'}, \qquad j = 1, 2, 3, 4. \end{cases} \tag{9.11}$$

where $\tilde{A}^j_{kj'}$'s and $\tilde{B}^j_{kj'}$'s denote the elements of the matrices $\tilde{A}_j$ and $\tilde{B}_j$ respectively.

As a result, values of $\tilde{\boldsymbol{b}}_0$, $\tilde{\boldsymbol{b}}_1$, $\tilde{\boldsymbol{b}}_2$, $\tilde{\boldsymbol{b}}_3$ and $\tilde{\boldsymbol{b}}_4$ are given by

$$\begin{aligned} \tilde{\boldsymbol{b}}_0 &= (3320, \widetilde{3777}, 3492)^T \\ \tilde{\boldsymbol{b}}_1 &= (765, 1023, \widetilde{545}, 1081)^T \\ \tilde{\boldsymbol{b}}_2 &= (2608, \widetilde{3284}, 2483, 2481, 1861, 2019, 1777)^T \\ \tilde{\boldsymbol{b}}_3 &= (2081, 3332, \widetilde{2218}, 2935, \widetilde{2554})^T \\ \tilde{\boldsymbol{b}}_4 &= (1569, 2240, 1598, 2038, \widetilde{1445}, 2374, 1493, \widetilde{1430})^T \end{aligned}$$

where the numbers with tildes are triangular fuzzy numbers.

For this numerical example, at each interaction with the DM, the corresponding augmented minimax problem (9.6) to obtain an α-Pareto optimal solution was solved through the genetic algorithm with decomposition procedures. The parameter values used in the genetic algorithm with decomposition procedures were set as follows: population size = 100, probability of crossover $p_c = 0.7$, probability of mutation $p_m = 0.005$, probability of inversion $p_i = 0.01$, $\varepsilon = 0.05$, $I_{\max} = 500$ and $I_{\min} = 200$.

First, the individual minimum and maximum of each objective function under the given constraints for $\alpha = 0$ and $\alpha = 1$ were calculated as shown in Table 9.1.

Table 9.1. Individual minimum and maximum for $\alpha = 0$ and 1

Objective function	$\alpha = 1$		$\alpha = 0$	
	Minimum	Maximum	Minimum	Maximum
$\boldsymbol{c}_1\boldsymbol{x}$	−5396.0	0.0	−5869.0	0.0
$\boldsymbol{c}_2\boldsymbol{x}$	0.0	6185.0	0.0	6682.0
$\boldsymbol{c}_3\boldsymbol{x}$	−4862.0	4367.0	−5036.0	4664.0

By taking account of the calculated individual minimum and maximum of each objective function for $\alpha = 0$ and $\alpha = 1$, the initial reference levels and the degree α were set to (−5396.0, 0.0, −4682.0; 1.0). For the initial reference levels and the degree α specified by the DM, the corresponding augmented minimax problem was solved, and the DM was supplied with the corresponding α-Pareto optimal solution and the objective function values shown in Interaction 1 of Table 9.2. On the basis of such information, since the DM was not satisfied with the current objective function values, the DM updated the reference levels and the degree α to $\bar{z}_1 = -2000.0$, $\bar{z}_2 = 0.0$, $\bar{z}_3 = -4862.0$ and $\alpha = 1.0$, respectively, to improve the values of $\boldsymbol{c}_2\boldsymbol{x}$ and $\boldsymbol{c}_3\boldsymbol{x}$ at the expense of $\boldsymbol{c}_1\boldsymbol{x}$. For the updated reference levels, the corresponding augmented minimax problem yielded the α-Pareto optimal solution and the objective function values as shown in Interaction 2 of Table 9.2. Since the DM was not satisfied with the α-Pareto optimal solution and objective function values, the DM updated (−2000.0, 0.0, −4862.0; 1.0) to (−2000.0, 0.0, −3000.0; 1.0) to improve the value of $\boldsymbol{c}_2\boldsymbol{x}$ at the expense of $\boldsymbol{c}_3\boldsymbol{x}$. The same procedure continues in this manner until the DM was satisfied with the current values of the objective functions. In this example, at the fourth interaction, the satisficing solution for the DM was derived. The whole interactive process is summarized in Table 9.2.

Table 9.2. Interaction process (20 trials)

Interaction	$\bar{z}_1$	$\bar{z}_2$	$\bar{z}_3$	α	$\boldsymbol{c}_1\boldsymbol{x}$	$\boldsymbol{c}_2\boldsymbol{x}$	$\boldsymbol{c}_3\boldsymbol{x}$	Number of solutions
1st	−5396.0	0.0	−4862.0	1.0	−3564.0	1919.0	−3267.0	20
2nd	−2000.0	0.0	−4862.0	1.0	−2477.0	1299.0	−3408.0	18
					−2257.0	1565.0	−3237.0	1
					−2297.0	1503.0	−3694.0	1
3rd	−2000.0	0.0	−3000.0	1.0	−1633.0	709.0	−2426.0	20
4th	−2000.0	0.0	−3000.0	0.3	−1703.0	623.0	−2923.0	20

9.3 Interactive fuzzy multiobjective 0-1 programming with fuzzy numbers

9.3.1 Fuzzy goals and solution concepts

For the α-multiobjective 0-1 programming problem (9.3) introduced in the previous section, considering the vague nature of DM's judgments, it is quite natural to assume that the DM may have a imprecise or fuzzy goal for each of the objective functions in the α-multiobjective 0-1 programming problem. In general, the fuzzy goals stated by the DM may be "$\boldsymbol{c}_i\boldsymbol{x}$ should be in the vicinity of some value" (called fuzzy equal) as well as "$\boldsymbol{c}_i\boldsymbol{x}$ should be substantially less than or equal to some value" (called fuzzy min) and "$\boldsymbol{c}_i\boldsymbol{x}$ should be substantially greater than or equal to some value" (called fuzzy max). Such a generalized α-multiobjective 0-1 programming problem may now be expressed as

$$\left.\begin{array}{r} \text{fuzzy min } z_i(\boldsymbol{x}) \; i \in I_1 \\ \text{fuzzy max } z_i(\boldsymbol{x}) \; i \in I_2 \\ \text{fuzzy equal } z_i(\boldsymbol{x}) \; i \in I_3 \\ \text{subject to } \boldsymbol{x} \in X(A, B, \boldsymbol{b}) \\ (A, B, \boldsymbol{b}, \boldsymbol{c}) \in (\tilde{A}, \tilde{B}, \tilde{\boldsymbol{b}}, \tilde{\boldsymbol{c}})_\alpha \end{array}\right\} \tag{9.12}$$

where $I_1 \cup I_2 \cup I_3 = \{1, 2, \ldots, k\}$, $I_i \cap I_j = \emptyset$, $i, j = 1, 2, 3$, $i \neq j$.

For this problem, M-α-Pareto optimal solutions can be defined.

Definition 9.3.1 (M-α-Pareto optimal solution).

A solution $\boldsymbol{x}^* \in X(A^*, B^*, \boldsymbol{b}^*)$ is said to be an M-α-Pareto optimal solution to the generalized α-multiobjective 0-1 programming problem, if and only if there does not exist another feasible solution $\boldsymbol{x} \in X(A, B, \boldsymbol{b})$, $(A, B, \boldsymbol{b}, \boldsymbol{c})$

$\in (\tilde{A}, \tilde{B}, \tilde{\boldsymbol{b}}, \tilde{\boldsymbol{c}})_\alpha$ *such that* $\mu_i(\boldsymbol{c}_i\boldsymbol{x}) \geq \mu_i(\boldsymbol{c}_i^*\boldsymbol{x}^*)$, $i = 1, \ldots, k$, *and* $\mu_j(\boldsymbol{c}_j\boldsymbol{x}) > \mu_j(\boldsymbol{c}_j^*\boldsymbol{x}^*)$ *for at least one* j $(1 \leq j \leq k)$. *The corresponding values of parameters* $(A^*, B^*, \boldsymbol{b}^*, \boldsymbol{c}^*)$ *are called M-*α*-level optimal parameters.*

Having elicited the membership functions $\mu_i(\boldsymbol{c}_i\boldsymbol{x})$, $i = 1, \ldots, k$, from the DM for each of the objective functions $\boldsymbol{c}_i\boldsymbol{x}$, $i = 1, \ldots, k$, the problem (9.12) can be written as the following fuzzy α-multiobjective decision making problem using the general aggregation function $\mu_D(\mu_1(\boldsymbol{c}_1\boldsymbol{x}), \ldots, \mu_k(\boldsymbol{c}_k\boldsymbol{x}))$:

$$\left.\begin{array}{ll} \text{maximize} & \mu_D(\mu_1(\boldsymbol{c}_1\boldsymbol{x}), \ldots, \mu_k(\boldsymbol{c}_k\boldsymbol{x})) \\ \text{subject to} & \boldsymbol{x} \in X(A, B, \boldsymbol{b}) \\ & (A, B, \boldsymbol{b}, \boldsymbol{c}) \in (\tilde{A}, \tilde{B}, \tilde{\boldsymbol{b}}, \tilde{\boldsymbol{c}})_\alpha \end{array}\right\} \quad (9.13)$$

where the value of the aggregation function $\mu_D(\boldsymbol{x})$ is interpreted as the degree of satisfaction of the DM for the whole k fuzzy goals [78].

However, as can be immediately understood from Definition 9.3.1, the M-α-Pareto optimal solutions to problem (9.13) consist of an infinite number of points and some subjective judgments should be added to the quantitative analyses by the DM. Namely, the DM must select a compromise or satisficing solution from the M-α-Pareto optimal solution set on the basis of DM's subjective value judgments.

9.3.2 Interactive fuzzy programming through genetic algorithms

To generate a candidate for the satisficing solution which is also an M-α-Pareto optimal solution, the DM is asked to specify the degree α of the α-level set and the aspiration levels of achievement for the membership values of all membership functions, called reference membership levels. For the DM's degree α and reference membership levels $\bar{\mu}_i$, $i = 1, \ldots, k$, the corresponding M-α-Pareto optimal solution, which is nearest to the requirement or better than if the reference membership levels are attainable in the minimax sense, is obtained by solving the following minimax problem in a membership function space [78]:

$$\left.\begin{array}{lllll} \text{minimize} & \displaystyle\max_{i=1,\ldots,k} \{\bar{\mu}_i - \mu_i(\boldsymbol{c}_i\boldsymbol{x})\} & & & \\ \text{subject to} & A_1\boldsymbol{x}_1 & + \cdots + & A_p\boldsymbol{x}_p & \leq \boldsymbol{b}_0 \\ & B_1\boldsymbol{x}_1 & & & \leq \boldsymbol{b}_1 \\ & & \ddots & & \vdots \\ & & & B_p\boldsymbol{x}_p & \leq \boldsymbol{b}_p \\ & \multicolumn{4}{l}{\boldsymbol{x}_j \in \{0, 1\}^{n_j}, \; j = 1, \ldots, p} \\ & \multicolumn{4}{l}{(A, B, \boldsymbol{b}, \boldsymbol{c}) \in (\tilde{A}, \tilde{B}, \tilde{\boldsymbol{b}}, \tilde{\boldsymbol{c}})_\alpha.} \end{array}\right\} \quad (9.14)$$

It should be emphasized here that, for generating M-α-Pareto optimal solutions by solving the minimax problem, if the uniqueness of the optimal solution is not guaranteed, it is necessary to perform the M-α-Pareto optimality test. To circumvent the necessity to perform the M-α-Pareto optimality test in the minimax problem (9.14), it is reasonable to use the following augmented minimax problem [78]:

$$\left.\begin{array}{ll} \text{minimize} & \max_{i=1,\ldots,k} \left\{ \bar{\mu}_i - \mu_i(\boldsymbol{c}_i \boldsymbol{x}) + \rho \sum_{i=1}^{k} (\bar{\mu}_i - \mu_i(\boldsymbol{c}_i \boldsymbol{x})) \right\} \\ \text{subject to} & A_1 \boldsymbol{x}_1 + \cdots + A_p \boldsymbol{x}_p \leq \boldsymbol{b}_0 \\ & B_1 \boldsymbol{x}_1 \qquad\qquad\quad \leq \boldsymbol{b}_1 \\ & \qquad \ddots \qquad\qquad\quad \vdots \\ & \qquad\qquad B_p \boldsymbol{x}_p \leq \boldsymbol{b}_p \\ & \boldsymbol{x}_j \in \{0,1\}^{n_j}, \ j = 1, \ldots, p \\ & (A, B, \boldsymbol{b}, \boldsymbol{c}) \in (\tilde{A}, \tilde{B}, \tilde{\boldsymbol{b}}, \tilde{\boldsymbol{c}})_\alpha \end{array}\right\} \tag{9.15}$$

where ρ is a sufficiently small positive number.

From the properties of the α-level sets for the matrices of the fuzzy numbers $\tilde{A}$ and $\tilde{B}$, and the vectors of the fuzzy numbers $\tilde{\boldsymbol{b}}$ and $\tilde{\boldsymbol{c}}$, the feasible regions for A, B, $\boldsymbol{b}$ and $\boldsymbol{c}$ can be represented respectively by the closed intervals $[A_\alpha^L, A_\alpha^R]$, $[B_\alpha^L, B_\alpha^R]$, $[\boldsymbol{b}_\alpha^L, \boldsymbol{b}_\alpha^R]$ and $[\boldsymbol{c}_\alpha^L, \boldsymbol{c}_\alpha^R]$.

For notational convenience, define

$$\mu_{iR}(\boldsymbol{c}_i \boldsymbol{x}) = \begin{cases} 1, & \boldsymbol{c}_i \boldsymbol{x} \leq z_i^1 \\ \mu_i(\boldsymbol{c}_i \boldsymbol{x}), & \boldsymbol{c}_i \boldsymbol{x} > z_i^1 \end{cases} \tag{9.16}$$

$$\mu_{iL}(\boldsymbol{c}_i \boldsymbol{x}) = \begin{cases} \mu_i(\boldsymbol{c}_i \boldsymbol{x}), & \boldsymbol{c}_i \boldsymbol{x} < z_i^1 \\ 1, & \boldsymbol{c}_i \boldsymbol{x} \geq z_i^1. \end{cases} \tag{9.17}$$

Figure 9.1 illustrates the the possible shapes of $\mu_{iR}(\cdot)$ and $\mu_{iL}(\cdot)$.

Observing that the DM chooses the most desirable parameter values $(A, B, \boldsymbol{b}, \boldsymbol{c}) \in (\tilde{A}, \tilde{B}, \tilde{\boldsymbol{b}}, \tilde{\boldsymbol{c}})_\alpha$, the value of the ith membership function can be defined as

$$M_i(\boldsymbol{x}) = \begin{cases} \mu_i(\boldsymbol{c}_i^L \boldsymbol{x}), & i \in I_1 \\ \mu_i(\boldsymbol{c}_i^R \boldsymbol{x}), & i \in I_2 \\ \min\{\mu_{iR}(\boldsymbol{c}_i^L \boldsymbol{x}), \mu_{iL}(\boldsymbol{c}_i^R \boldsymbol{x})\}, & i \in I_3 \end{cases} \tag{9.18}$$

Then, the problem (9.15) can be rewritten as follows:

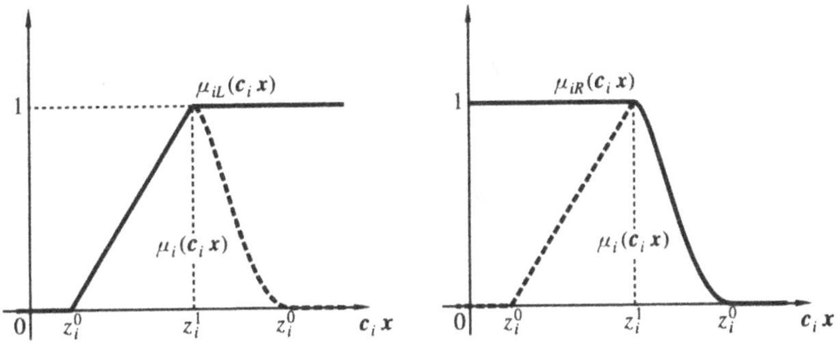

Figure 9.1. $\mu_{iL}(\cdot)$ and $\mu_{iR}(\cdot)$.

$$\left.\begin{array}{ll}\text{minimize} & \max_{i=1,\ldots,k}\left\{(\bar{\mu}_i - M_i(\boldsymbol{x})) + \rho\sum_{i=1}^{k}(\bar{\mu}_i - M_i(\boldsymbol{x}))\right\} \\ \text{subject to} & A_{1\alpha}^L\boldsymbol{x}_1 + \cdots + A_{p\alpha}^L\boldsymbol{x}_p \leq \boldsymbol{b}_{0\alpha}^R \\ & B_{1\alpha}^L\boldsymbol{x}_1 \leq \boldsymbol{b}_{1\alpha}^R \\ & \quad \ddots \qquad \vdots \\ & B_{p\alpha}^L\boldsymbol{x}_p \leq \boldsymbol{b}_{p\alpha}^R \\ & \boldsymbol{x}_j \in \{0,1\}^{n_j}, \ j = 1,\ldots,p. \end{array}\right\} \tag{9.19}$$

Since the constraints of the problem (9.19) preserve the block angular structure, the genetic algorithm with decomposition procedures explained in Chapter 7 is applicable.

Note that the fitness of an individual **S** is defined as

$$f(\mathbf{S}) = (1.0 + k\rho) - \max_{i=1,\ldots,k}\left\{(\bar{\mu}_i - M_i(\boldsymbol{x})) + \rho\sum_{i=1}^{k}(\bar{\mu}_i - M_i(\boldsymbol{x}))\right\} \tag{9.20}$$

where $\boldsymbol{x}$ denotes a solution (phenotype) generated from the individual (genotype) **S** by the decoding algorithm.

We can now construct the interactive algorithm in order to derive the satisficing solution for the DM from the M-α-Pareto optimal solution set.

Interactive fuzzy multiobjective 0-1 programming with fuzzy numbers

Step 0: Calculate the individual minimum and maximum of each objective function under the given constraints for $\alpha = 0$ and $\alpha = 1$.

Step 1*: Elicit a membership function from the DM for each of the objective functions by taking account of the calculated individual minimum and maximum of each objective function for $\alpha = 0$ and $\alpha = 1$. Then ask the DM to select the initial value of α $(0 \le \alpha \le 1)$ and the initial reference membership levels $\bar{\mu}_i$, $i = 1, \ldots, k$.

Step 2: Set an iteration index (generation) $t = 0$ and determine the parameter values for the convergence criterion ε, the minimal search generation $I_{\min}$, the maximal search generation $I_{\max}$, the probability of crossover p_c, the probability of mutation p_m, the probability of inversion p_i.

Step 3: Generate N individuals of the triple string type at random as the initial population.

Step 4: Evaluate each individual (subindividual) on the basis of the phenotype obtained by the proposed decoding algorithm and calculate the mean fitness f_{mean} and the maximal fitness $f_{\max}$ of the population. If $t > I_{\min}$ and $(f_{\max} - f_{\text{mean}})/f_{\max} < \varepsilon$, or, if $t > I_{\max}$, regard the individual with the maximal fitness as the best individual, and stop. Otherwise, set $t = t + 1$ and proceed to Step 5.

Step 5†: Apply the reproduction operator to every subindividual.

Step 6†: Apply the PMX for double strings to the middle and lower part of every subindividual according to the probability of crossover p_c.

Step 7†: Apply the mutation operator of the bit-reverse type to the lower part of every subindividual according to the probability of mutation p_m, and apply the inversion operator for the middle and lower part of every subindividual according to the probability of inversion p_i.

Step 8: Apply the PMX for upper strings according to p_c.

Step 9: Apply the inversion operator for upper strings according to p_i and return to Step 4.

Step 10*: If the DM is satisfied with the current values of the membership functions and/or objective functions given by the current best solution, stop. Otherwise, ask the DM to update the reference membership levels and α by taking into account the current values of membership functions and/or objective functions, and return to Step 3.

Observe that the operations in the steps marked with † can be applied to every subindividual independently. For this reason, it is theoretically possible to reduce the amount of working memory and carry out parallel processing.

9.3.3 Numerical experiments

For investigating the feasibility and efficiency of the proposed method, consider the following multiobjective 0-1 programming problem with the block angular structure involving fuzzy numbers (3 objectives, 3 blocks, 20 (=4+9+7) variables and 3 coupling constraints):

$$
\left.\begin{array}{rllll}
\text{minimize} & \tilde{\boldsymbol{c}}_{11}\boldsymbol{x}_1 & + \ \tilde{\boldsymbol{c}}_{12}\boldsymbol{x}_2 & + \ \tilde{\boldsymbol{c}}_{13}\boldsymbol{x}_3 & \\
\text{minimize} & \tilde{\boldsymbol{c}}_{21}\boldsymbol{x}_1 & + \ \tilde{\boldsymbol{c}}_{22}\boldsymbol{x}_2 & + \ \tilde{\boldsymbol{c}}_{23}\boldsymbol{x}_3 & \\
\text{minimize} & \tilde{\boldsymbol{c}}_{31}\boldsymbol{x}_1 & + \ \tilde{\boldsymbol{c}}_{32}\boldsymbol{x}_2 & + \ \tilde{\boldsymbol{c}}_{33}\boldsymbol{x}_3 & \\
\text{subject to} & \tilde{A}_1\boldsymbol{x}_1 & + \ \tilde{A}_2\boldsymbol{x}_2 & + \ \tilde{A}_3\boldsymbol{x}_3 & \leq \tilde{\boldsymbol{b}}_0 \\
 & \tilde{B}_1\boldsymbol{x}_1 & & & \leq \tilde{\boldsymbol{b}}_1 \\
 & & \tilde{B}_2\boldsymbol{x}_2 & & \leq \tilde{\boldsymbol{b}}_2 \\
 & & & \tilde{B}_3\boldsymbol{x}_3 & \leq \tilde{\boldsymbol{b}}_3 \\
 & \multicolumn{4}{c}{\boldsymbol{x}_j \in \{0,1\}^{n_j}, \ j = 1,2,3.}
\end{array}\right\} \quad (9.21)
$$

As a numerical example with 20 variables, we used the following values of $\tilde{\boldsymbol{c}}_i = (\tilde{\boldsymbol{c}}_{i1}, \tilde{\boldsymbol{c}}_{i2}, \tilde{\boldsymbol{c}}_{i3})$, $i = 1,2,3$, $\tilde{A}_j$ and $\tilde{B}_j$, $j = 1,2,3$. Observe that the elements of $\tilde{\boldsymbol{c}}_1$, $\tilde{\boldsymbol{c}}_2$ and $\tilde{\boldsymbol{c}}_3$ were randomly determined from the closed intervals $[-999, 0]$, $[0, 999]$ and $[-999, 999]$, respectively, while the elements of $\tilde{A}_j$ and $\tilde{B}_j$, $j = 1,2,3$, were randomly determined from the closed interval $[0, 999]$, where the numbers with tildes are the triangular fuzzy numbers.

$$
\begin{array}{rl}
\tilde{\boldsymbol{c}}_1 = (& -615, \ -172, \ -\widetilde{554}, \ -292, \ -872, \ -835, \ -844, \ -895, \ -594, \ -540, \\
 & -168, \ -654, \ -\widetilde{690}, \ -263, \ -106, \ -814, \ -191, \ -423, \ -351, \ -839 \ ) \\
\tilde{\boldsymbol{c}}_2 = (& 519, \ 607, \ 732, \ 556, \ 344, \ \widetilde{801}, \ \widetilde{590}, \ \widetilde{266}, \ 670, \ 552, \\
 & 788, \ 887, \ 889, \ 68, \ 800, \ 907, \ 644, \ \widetilde{165}, \ 301, \ 166 \ ) \\
\tilde{\boldsymbol{c}}_3 = (& -153, \ 33, \ -378, \ -20, \ 915, \ 729, \ -982, \ 112, \ -\widetilde{799}, \ 738, \\
 & 739, \ 362, \ 183, \ 115, \ \widetilde{788}, \ -790, \ \widetilde{402}, \ -182, \ -331, \ 964 \ )
\end{array}
$$

$$
\tilde{A}_1 = \begin{pmatrix} 865 & \widetilde{148} & 171 & 68 \\ 651 & 736 & 102 & 160 \\ 93 & 121 & 24 & \widetilde{762} \end{pmatrix}, \quad
\tilde{A}_3 = \begin{pmatrix} 44 & 882 & 108 & 520 & \widetilde{578} & 9 & 387 \\ 775 & 354 & 697 & 912 & 670 & \widetilde{705} & 426 \\ 502 & 194 & 644 & 127 & 264 & 336 & 703 \end{pmatrix},
$$

$$
\tilde{A}_2 = \begin{pmatrix} 956 & 27 & 646 & 108 & 427 & 309 & 18 & \widetilde{885} & 757 \\ 509 & 165 & \widetilde{762} & 880 & 499 & 875 & 734 & 235 & 51 \\ 605 & 875 & 504 & 678 & 989 & \widetilde{604} & 496 & 589 & 895 \end{pmatrix},
$$

$$
\tilde{B}_1 = \begin{pmatrix} 513 & \widetilde{175} & 308 & 534 \\ 947 & 171 & \widetilde{702} & 226 \\ \widetilde{494} & 124 & 83 & 389 \\ 277 & 368 & 983 & \widetilde{535} \end{pmatrix}, \quad
\tilde{B}_3 = \begin{pmatrix} 400 & 773 & 244 & 342 & 229 & 297 & 304 \\ \widetilde{398} & 676 & 732 & 937 & 233 & 838 & 967 \\ 809 & 158 & \widetilde{279} & 135 & 864 & 750 & 207 \\ 218 & \widetilde{563} & 715 & 197 & 989 & 250 & 430 \\ 978 & 395 & 432 & 127 & \widetilde{457} & 237 & 986 \end{pmatrix},
$$

$$\tilde{B}_2 = \begin{pmatrix} 765 & 644 & 767 & \widetilde{780} & 822 & 151 & 625 & 314 & 346 \\ 917 & 519 & 401 & 606 & 785 & 931 & \widetilde{869} & 866 & 674 \\ \widetilde{758} & 581 & 389 & 355 & 200 & 826 & 415 & 463 & 979 \\ 126 & 212 & 958 & 737 & 409 & \widetilde{780} & 757 & 956 & 28 \\ 318 & 756 & \widetilde{242} & 589 & 43 & 956 & 319 & 59 & 441 \\ 915 & \widetilde{572} & 118 & 569 & 252 & 495 & 236 & 476 & 406 \\ 872 & 426 & 358 & 381 & 43 & 160 & 522 & \widetilde{696} & 97 \end{pmatrix}$$

Each element $\tilde{b}_j(k)$ of $\tilde{\boldsymbol{b}}_j$ was determined by

$$\begin{cases} \tilde{b}_0(k) = 0.5 \times \sum_{j=1}^{3} \sum_{j'=1}^{n_j} \tilde{A}^j_{kj'} \\ \tilde{b}_j(k) = 0.5 \times \sum_{j'=1}^{n_j} \tilde{B}^j_{kj'}, \quad j = 1, 2, 3 \end{cases} \tag{9.22}$$

where $\tilde{A}^j_{kj'}$'s and $\tilde{B}^j_{kj'}$'s denote the elements of the matrices $\tilde{A}_j$ and $\tilde{B}_j$ respectively.

As a result, values of $\tilde{\boldsymbol{b}}_0$ were given by

$$\begin{aligned} \tilde{\boldsymbol{b}}_0 &= (3956.5, \widetilde{5449.0}, 5002.5)^T \\ \tilde{\boldsymbol{b}}_1 &= (765.0, 1023.0, \widetilde{545.0}, \widetilde{1081.5})^T \\ \tilde{\boldsymbol{b}}_2 &= (2608.0, \widetilde{3284.0}, 2483.0, 2481.5, \widetilde{1861.5}, 2019.5, 1777.5)^T \\ \tilde{\boldsymbol{b}}_3 &= (1294.5, 2390.5, \widetilde{1601.0}, 1681.0, 1806.0)^T \end{aligned}$$

where the numbers with tildes are triangular fuzzy numbers.

Concerning the fuzzy goals of the decision maker (DM), the values of z_i^1 and z_i^0 were set in the same way as adopted by Zimmermann [142].

For this numerical example, at each interaction with the DM, the corresponding augmented minimax problem was solved through 20 trials of the genetic algorithm with decomposition procedures. The parameter values used in the genetic algorithm with decomposition procedures were set as follows: population size = 50, probability of crossover $p_c = 0.7$, probability of mutation $p_m = 0.005$, probability of inversion $p_i = 0.01$, $\varepsilon = 0.05$, $I_{\max} = 1000$ and $I_{\min} = 300$.

The augmented minimax problem (9.19) was solved for the degree $\alpha = 1.0$ specified by the DM and the initial reference membership levels 1.0, and the DM was supplied with the corresponding α-Pareto optimal solution and both membership function values and objective function values shown in the first interaction of Table 9.3. On the basis of such information, since the DM was not satisfied with the current membership function values or objective function values, the DM updated the reference membership levels to $\bar{\mu}_1 = 0.8$, $\bar{\mu}_2 = 1.0$, $\bar{\mu}_3 = 1.0$ and $\alpha = 1.0$ in order to improve the satisfaction levels for μ_2 and μ_3 at the expense of μ_1. For the updated reference membership levels, the corresponding augmented minimax problem yielded the α-Pareto optimal solution and both membership function values and objective function values shown in the second interaction of Table 9.3. The same procedure continues in this manner until the DM was satisfied with the current values of the membership functions and the objective functions. In this example, at the fourth interaction, the satisficing solution for the DM was derived. The whole interaction process is summarized in Table 9.3.

Table 9.3. Interaction process

Interaction $(\bar{\mu}_1, \bar{\mu}_2, \bar{\mu}_3, \alpha)$	$c_1 x$	$c_2 x$	$c_3 x$	$\mu_1(c_1 x)$	$\mu_2(c_2 x)$	$\mu_3(c_3 x)$	Number of solutions
1st	−2948	2045	−1822	0.5572	0.5703	0.5942	18
(1.0, 1.0, 1.0, 1.0)	−2947	1895	−1885	0.5570	0.6018	0.6117	2
2nd	−2684	1827	−2000	0.5073	0.6161	0.6438	19
(0.8, 1.0, 1.0, 1.0)	−2948	2045	−1822	0.5572	0.5703	0.5942	1
3rd	−2756	1691	−1851	0.5209	0.6447	0.6023	20
(0.8, 1.0, 0.9, 1.0)							
4th	−2756	1579	−1891	0.5209	0.6682	0.6134	20
(0.8, 1.0, 0.9, 0.8)							

Further computational experiments were performed on numerical examples with 30, 50, 70 and 100 variables through the genetic algorithm with decomposition procedures (GADP). For comparison, the same numerical examples were solved through the genetic algorithm with double string (GADS) [82, 96, 97, 98, 106, 107, 109].

The computational experiments were performed on a personal computer (processor: Pentium-120MHz, memory: 48MB, OS: Windows NT 4.0), where, at each interaction with the DM, the corresponding augmented minimax problem was solved through both GADP and GADS (5 trials). The parameter values used in both GADP and GADS were: population size = 100, probability of crossover $p_c = 0.7$, probability of mutation $p_m = 0.005$, probability of inversion $p_i = 0.01$, $\varepsilon = 0.05$, $I_{\max} = 1000$ and $I_{\min} = 200$.

First, consider another multiobjective 0-1 programming problem with a block angular structure involving fuzzy numbers, which has 3 objectives, 3 blocks, 30 (=12+8+10) variables and 3 coupling constraints. Tables 9.4 and

9.5 show the interaction processes with the DM through GADP and GADS respectively.

Table 9.4. Interaction process (GADP, 5 trials)

Interaction $(\bar{\mu}_1, \bar{\mu}_2, \bar{\mu}_3, \alpha)$	c_1x	c_2x	c_3x	$\mu_1(c_1x)$	$\mu_2(c_2x)$	$\mu_3(c_3x)$	Number of solutions
1st	−6178	2916	−71	0.6525	0.6737	0.6554	4
(1.0, 1.0, 1.0, 1.0)	−6179	3217	−215	0.6526	0.6486	0.6723	1
(Optimum)	−6178	2916	−71	0.6525	0.6737	0.6554	
2nd	−4892	2123	−840	0.5356	0.7398	0.7459	5
(0.8, 1.0, 1.0, 1.0)							
(Optimum)	−4892	2123	−840	0.5356	0.7398	0.7459	
3rd	−4979	2143	−682	0.5435	0.7381	0.7273	5
(0.8, 1.0, 0.9, 1.0)							
(Optimum)	−4979	2143	−682	0.5435	0.7381	0.7273	
4th	−4989.9	2084.1	−27.5	0.5445	0.7430	0.6503	5
(0.8, 1.0, 0.9, 0.7)							
(Optimum)	−4989.9	2084.1	−27.5	0.5445	0.7430	0.6503	

Table 9.5. Interaction process (GADS, 5 trials)

Interaction $(\bar{\mu}_1, \bar{\mu}_2, \bar{\mu}_3, \alpha)$	c_1x	c_2x	c_3x	$\mu_1(c_1x)$	$\mu_2(c_2x)$	$\mu_3(c_3x)$	Number of solutions
1st	−6178	2916	−71	0.6525	0.6737	0.6554	4
(1.0, 1.0, 1.0, 1.0)	−6179	3217	−215	0.6526	0.6486	0.6723	1
(Optimum)	−6178	2916	−71	0.6525	0.6737	0.6554	
2nd	−4892	2123	−840	0.5356	0.7398	0.7459	5
(0.8, 1.0, 1.0, 1.0)							
(Optimum)	−4892	2123	−840	0.5356	0.7398	0.7459	
3rd	−4979	2143	−682	0.5435	0.7381	0.7273	5
(0.8, 1.0, 0.9, 1.0)							
(Optimum)	−4979	2143	−682	0.5435	0.7381	0.7273	
4 th	−4989.9	2084.1	−27.5	0.5445	0.7430	0.6503	5
(0.8, 1.0, 0.9, 0.7)							
(Optimum)	−4989.9	2084.1	−27.5	0.5445	0.7430	0.6503	

In the first interaction, the augmented minimax problem was solved for the degree $\alpha = 1.0$ specified by the DM and the initial reference membership levels 1.0, and the DM was supplied with the corresponding M-α-Pareto optimal solution and both membership function values and objective function values shown in the first interaction of Tables 9.4 and 9.5. On the basis of such information, since the DM was not satisfied with the current membership function values or objective function values, the DM updated the reference membership levels to $\bar{\mu}_1 = 0.8$, $\bar{\mu}_2 = 1.0$, $\bar{\mu}_3 = 1.0$ and $\alpha = 1.0$ in order to improve the satisfaction levels for $\mu_2(\boldsymbol{c}_2\boldsymbol{x})$ and $\mu_3(\boldsymbol{c}_3\boldsymbol{x})$ at the expense of $\mu_1(\boldsymbol{c}_1\boldsymbol{x})$. For the updated reference membership levels, the corresponding augmented minimax problem yielded the M-α-Pareto optimal solution and both membership function values and objective function values shown in the second interaction of Tables 9.4 and 9.5. The same procedure continues in this manner until the DM was satisfied with the current values of the membership functions and the objective functions. In this example, as shown in Tables 9.4 and 9.5, at the fourth interaction, a satisficing solution for the DM was derived.

In this example, the average processing time (APT) through all interaction processes and the average relative error (ARE)

$$\text{ARE} = \left| \frac{\text{Best value obtained by GA} - \text{Exact optimal value}}{\text{Exact optimal value}} \right| \tag{9.23}$$

are shown in Table 9.6. From Table 9.6, for the problem with 30 variables, it can be seen that the average processing time (APT) of GADP is lower than that of GADS.

Table 9.6. Average processing time (APT) and average relative error (ARE) for a problem with 30 variables

	APT (sec)	ARE (×100%)
GADP	211.010	0.1122
GADS	242.654	0.1122

Next, consider another multiobjective 0-1 programming problem with the block angular structure involving fuzzy numbers, which has 3 objectives, 5 blocks, 50 (=12+8+10+7+13) variables and 5 coupling constraints.

Table 9.7 shows the average processing time (APT) and the average relative error (ARE) for this problem. From Table 9.7, for the problem with 50 variables, it can be recognized that GADP gives better results than GADS, both for APT and ARE.

To see how the processing times of both GADS and GADP change with the size of the problem, similar computational experiences were performed on

Table 9.7. Average processing time (APT) and average relative error (ARE) for a problem with 50 variables

	APT (sec)	ARE (×100%)
GADP	315.366	1.299
GADS	383.333	1.472

numerical examples with 70 and 100 variables. Figure 9.2 shows processing times for typical problems with 30, 50, 70 and 100 variables.

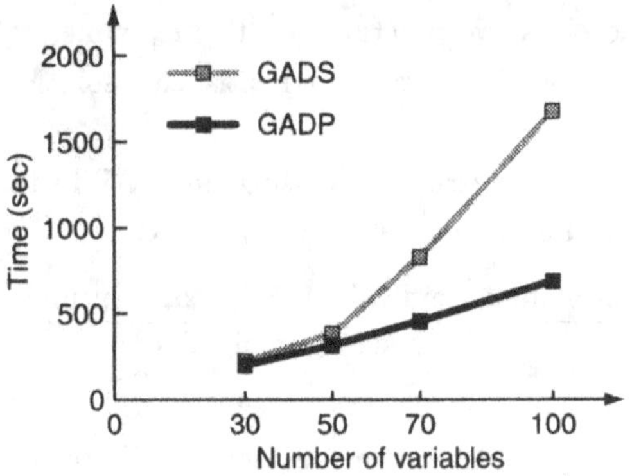

Figure 9.2. Processing time for typical problems with 30, 50, 70 and 100 variables (GADS versus GADP).

From our limited computational experiments through GADS and GADP, as an approximate solution method for multidimensional 0-1 knapsack problems with block angular structures involving fuzzy numbers, GADP seems to be more efficient and effective than GADS.

It is interesting to note here that GADP has recently been improved through the use of the linear programming relaxation. Interested readers might refer to Kato and Sakawa [53] for details. However, further extensions of GADP not only to more general large scale 0-1 programming problems but also to large scale integer programming problems would be required.

9.4 Conclusion

In this chapter, we focused on the large scale multiobjective 0-1 knapsack problem with the block angular structure involving fuzzy numbers, and introduced the α-Pareto optimality as an extended Pareto optimality concept. In our interactive decision making method, for the degree α and the reference levels specified by the DM, the corresponding α-Pareto optimal solution can be obtained by solving the augmented minimax problems through the genetic algorithms with decomposition procedures. By updating the reference levels and/or the degree α, a satisficing solution for the DM can be derived efficiently from an α-Pareto optimal solution set. An illustrative numerical example demonstrated the feasibility and efficiency of the proposed method.

Furthermore, by incorporating the fuzzy goals of the DM together with the M-α-Pareto optimality, if the DM specifies the degree α and the reference membership levels, the corresponding M-α-Pareto optimal solution can be obtained by solving the augmented minimax problems for which the genetic algorithm with decomposition procedures is applicable. Then we presented an interactive fuzzy satisficing method for deriving a satisficing solution for the DM efficiently from an M-α-Pareto optimal solution set by updating the reference membership levels and/or the degree α. Through some computational experiments on numerical examples with 20, 30, 50, 70 and 100 variables, both feasibility and efficiency of the proposed method were demonstrated. However, applications of the proposed method as well as extensions to more general cases will be desirable in the future.

10. Further Research Directions

In the preceding chapters, on the basis of the author's continuing research works, we have discussed the latest advances in the new field of interactive multiobjective optimization for large scale linear and 0-1 programming problems with the block angular under fuzziness. However, many other interesting related topics, including multiobjective linear fractional and nonlinear programming problems with block angular structures under fuzziness, still remain to be further explored. In this chapter, being limited by space, we only outline the latest major research results of these topics because of their close relationship to our discussions in the preceding chapters.

10.1 Large scale fuzzy multiobjective linear fractional programming

10.1.1 Introduction

In 1962, a linear fractional programming problem was first formulated by Charnes and Cooper [7] as an optimization problem of a ratio of linear functions subject to linear constraints. As indicated in Kornbluth and Steuer [55], linear fractional objectives (i.e., ratio objectives that have linear numerator and denominator) are useful in production planning, financial and corporate planning, health care, hospital planning and so forth. Examples of fractional objectives in production planning include inventory/sales, output/employee, etc. For single objective linear fractional programming, the Charnes and Cooper [7] transformation can be used to transform the problem into a linear programming problem. Since then, single objective linear fractional programming has been extensively studied as can be seen in the invited review article of Schaible and Ibaraki [125]. However, concerning multiobjective linear fractional programming (MOLFP), few approaches have appeared in the literature [9, 54, 55, 63, 117, 111]. Kornbluth and Steuer [54, 55] presented two different approaches to MOLFP based on the weighted Tchebycheff norm. Luhandjula [63] proposed a linguistic approach to multiobjective linear fractional programming by introducing linguistic variables to represent linguistic aspirations of the decision maker (DM). In the framework of the fuzzy decision of Bellman and Zadeh [3], Sakawa and Yumine [111] presented a fuzzy programming approach for solving MOLFP by combined use of the bisection method and the linear programming method. As a generalization of the result in Sakawa and Yumine [111], Sakawa and Yano [117] presented a linear programming-based interactive fuzzy satisficing method for multiobjective linear fractional programming to derive the satisficing solution for the DM efficiently from a Pareto optimal solution set by updating reference membership levels.

However, it is significant to realize here that many decision making problems arising in application are often formulated as mathematical programming problems with special structures that can be exploited. One familiar structure is the block angular structure to the constraints. At this juncture, decomposition methods have several significant advantages over conventional methods in solving single- and multiple-objective programming problems with the block angular structure [13, 14, 34, 58, 137, 104].

From such a point of view, as discussed in Chapters 4 and 5, we have already presented fuzzy programming approaches to single and multiple linear programming problems with the block angular structure by incorporating the Dantzig-Wolfe decomposition method [103, 81]. These results are immediately extended to an interactive fuzzy satisficing method [102].

In this section, we focus on MOLFP problems with block angular structures. Considering the vague nature of human judgments, it is quite natural to assume that the DM may have a fuzzy goal for each of the objective functions. Through the interaction with the DM, these fuzzy goals can be quantified by eliciting the corresponding membership functions including nonlinear ones. Then a linear programming-based interactive decision making method with decomposition procedures for deriving a satisficing solution for the DM efficiently from a Pareto optimal solution set is presented. In our interactive decision making method, for the reference membership levels specified by the DM, the corresponding Pareto optimal solution can be easily obtained by solving the minimax problems for which the Dantzig-Wolfe decomposition method [13, 14] and Ritter's partitioning procedure [72] are applicable. The satisficing solution for the DM can be derived efficiently from Pareto optimal solutions by updating the reference membership levels based on the current values of the Pareto optimal solution together with the trade-off information between the membership functions.

10.1.2 Problem formulation

As a linear fractional generalization of Chapter 5, consider a multiobjective linear fractional programming problem with the following block angular structure:

$$\left.\begin{array}{ll} \text{minimize} & z_1(\boldsymbol{x}, \boldsymbol{c}_1, \boldsymbol{d}_1) \\ & \vdots \\ \text{minimize} & z_k(\boldsymbol{x}, \boldsymbol{c}_k, \boldsymbol{d}_k) \\ \text{subject to} & A_1\boldsymbol{x}_1 + \cdots + A_p\boldsymbol{x}_p \leq \boldsymbol{b}_0 \\ & B_1\boldsymbol{x}_1 \qquad\qquad\qquad \leq \boldsymbol{b}_1 \\ & \qquad\qquad \ddots \qquad\qquad \vdots \\ & \qquad\qquad\qquad B_p\boldsymbol{x}_p \leq \boldsymbol{b}_p \\ & \boldsymbol{x}_j \geq 0,\ j = 1, \ldots, p, \end{array}\right\} \tag{10.1}$$

where $\boldsymbol{x}_j$, $j = 1, \ldots, p$, are n_j dimensional column vectors of decision variables, $A_1\boldsymbol{x}_1 + \cdots + A_p\boldsymbol{x}_p \leq \boldsymbol{b}_0$ are m_0 dimensional coupling constraints, A_j, $j = 1, \ldots, p$, are $m_0 \times n_j$ coefficient matrices. $B_j\boldsymbol{x}_j \leq \boldsymbol{b}_j$ are m_j dimensional block constraints with respect to $\boldsymbol{x}_j$, B_j, $j = 1, \ldots, p$, are $m_j \times n_j$ coefficient matrices and $\boldsymbol{b}_j$, $j = 1, \ldots, p$, are m_j dimensional column vectors.

Furthermore, $z_1(\boldsymbol{x}, \boldsymbol{c}_1, \boldsymbol{d}_1), \ldots, z_k(\boldsymbol{x}, \boldsymbol{c}_k, \boldsymbol{d}_k)$ are k distinct linear fractional objective functions defined by

$$z_i(\boldsymbol{x}, \boldsymbol{c}_i, \boldsymbol{d}_i) = \frac{p_i(\boldsymbol{x}, \boldsymbol{c}_i)}{q_i(\boldsymbol{x}, \boldsymbol{d}_i)} = \frac{\boldsymbol{c}_{i1}\boldsymbol{x}_1 + \cdots + \boldsymbol{c}_{ip}\boldsymbol{x}_p + c_{i,p+1}}{\boldsymbol{d}_{i1}\boldsymbol{x}_1 + \cdots + \boldsymbol{d}_{ip}\boldsymbol{x}_p + d_{i,p+1}}, \tag{10.2}$$

where $\boldsymbol{x} = (\boldsymbol{x}_1^T, \ldots, \boldsymbol{x}_p^T)^T$, is a $\sum_{j=1}^p n_j$ dimensional vector, $\boldsymbol{c}_i = (\boldsymbol{c}_{i1}, \ldots, \boldsymbol{c}_{ip}, c_{i,p+1})$ and $\boldsymbol{d}_i = (\boldsymbol{d}_{i1}, \ldots, \boldsymbol{d}_{ip}, d_{i,p+1})$, $i = 1, \ldots, k$, are $(\sum_{i=1}^p n_i + 1)$ dimensional row vectors. Here it is customary to assume that the $q_i(\boldsymbol{x}, \boldsymbol{d}_i) > 0$, $i = 1, \ldots, k$, for all $\boldsymbol{x} \in X$, where X denotes the feasible region satisfying all of the constraints of the problem (10.1). In the followings, for notational convenience, define the following matrices and vectors.

$$A = (A_1, \ldots, A_p), \quad \boldsymbol{b} = \left(\boldsymbol{b}_0^T, \boldsymbol{b}_1^T, \ldots, \boldsymbol{b}_p^T\right)^T$$

$$B = \begin{bmatrix} B_1 & & O \\ & \ddots & \\ O & & B_p \end{bmatrix}$$

$$\boldsymbol{c} = (\boldsymbol{c}_1, \ldots, \boldsymbol{c}_k), \quad \boldsymbol{d} = (\boldsymbol{d}_1, \ldots, \boldsymbol{d}_k)$$

In general, however, for multiobjective programming problems, a complete optimal solution which simultaneously minimizes all of the multiple objective functions does not always exist when the objective functions conflict with each other. Thus, instead of a complete optimal solution, Pareto optimality is introduced in multiobjective programming problems [78].

However, considering the imprecise nature of the DM's judgments, it is reasonable to assume that the DM may have a fuzzy goal for each of the objective functions in the multiobjective linear fractional programming problem (10.1). For example, a goal stated by the DM may be to achieve "substantially less than or equal to p_i". This type of statement can be quantified by eliciting a corresponding membership function.

In order to elicit a membership function $\mu_i(z_i(\boldsymbol{x}, \boldsymbol{c}_i, \boldsymbol{d}_i))$ from the DM for each of the objective functions $z_i(\boldsymbol{x}, \boldsymbol{c}_i, \boldsymbol{d}_i)$, $i = 1, \ldots, k$, in the problem (10.1), we first calculate the individual minimum and maximum of each objective function under the given constraints. By taking account of the calculated individual minimum and maximum of each objective function together with the rate of increase of membership of satisfaction, the DM must determine a membership function $\mu_i(z_i(\boldsymbol{x}, \boldsymbol{c}_i, \boldsymbol{d}_i))$ in a subjective manner.

It is significant to note here that, in the fuzzy approaches, we can treat two types of fuzzy goals, namely, fuzzy goals expressed in words such as "$z_i(\boldsymbol{x}, \boldsymbol{c}_i, \boldsymbol{d}_i)$ should be in the vicinity of r_i" (called fuzzy equal) as well as "$z_i(\boldsymbol{x}, \boldsymbol{c}_i, \boldsymbol{d}_i)$ should be substantially less than or equal to p_i or greater than or equal to q_i" (called fuzzy min or fuzzy max). Such a generalized multiobjective linear fractional programming problem can be expressed as

$$\left.\begin{array}{lll}
\text{fuzzy min} & z_i(\boldsymbol{x}, \boldsymbol{c}_i, \boldsymbol{d}_i), & i \in I_1 \\
\text{fuzzy max} & z_i(\boldsymbol{x}, \boldsymbol{c}_i, \boldsymbol{d}_i), & i \in I_2 \\
\text{fuzzy equal} & z_i(\boldsymbol{x}, \boldsymbol{c}_i, \boldsymbol{d}_i), & i \in I_3 \\
\text{subject to} & A_1\boldsymbol{x}_1 + \cdots + A_p\boldsymbol{x}_p & \leq \boldsymbol{b}_0 \\
 & B_1\boldsymbol{x}_1 & \leq \boldsymbol{b}_1 \\
 & \quad \ddots & \vdots \\
 & \qquad B_p\boldsymbol{x}_p & \leq \boldsymbol{b}_p \\
 & \boldsymbol{x}_j \geq \boldsymbol{0}, \; j = 1, \ldots, p &
\end{array}\right\} \tag{10.3}$$

where $I_1 \cup I_2 \cup I_3 = \{1, 2, \ldots, k\}$, $I_i \cap I_j = \emptyset$, $i = 1, 2, 3$, $i \neq j$.

In order to elicit a membership functions $\mu_i(z_i(\boldsymbol{x}, \boldsymbol{c}_i, \boldsymbol{d}_i))$ from the DM for a fuzzy goal like "$z_i(\boldsymbol{x}, \boldsymbol{c}_i, \boldsymbol{d}_i)$ should be in the vicinity of r_i", it is obvious that we can use different functions to the left and right side of r_i. For the membership functions of the problem (10.3), it is reasonable to assume that $\mu_i(z_i(\boldsymbol{x}, \boldsymbol{c}_i, \boldsymbol{d}_i))$, $i \in I_1$ and the right side functions of $\mu_i(z_i(\boldsymbol{x}, \boldsymbol{c}_i, \boldsymbol{d}_i))$, $i \in I_3$ are strictly monotone decreasing and continuous functions with respect to $z_i(\boldsymbol{x}, \boldsymbol{c}_i, \boldsymbol{d}_i)$, and $\mu_i(z_i(\boldsymbol{x}, \boldsymbol{c}_i, \boldsymbol{d}_i))$, $i \in I_2$ and the left side functions of $\mu_i(z_i(\boldsymbol{x}, \boldsymbol{c}_i, \boldsymbol{d}_i))$, $i \in I_3$ are strictly monotone increasing and continuous with respect to $z_i(\boldsymbol{x}, \boldsymbol{c}_i, \boldsymbol{d}_i)$.

When fuzzy equal is included in the fuzzy goals of the DM, it is desirable that $z_i(\boldsymbol{x}, \boldsymbol{c}_i, \boldsymbol{d}_i)$ should be as close to r_i as possible. Consequently, the notion of Pareto optimal solutions defined in terms of objective functions cannot be applied. For this reason, the concept of M-Pareto optimal solutions which is defined in terms of membership functions instead of objective functions has been introduced [78].

However, as can be immediately understood from the definition, in general, M-Pareto optimal solutions to the problem (10.3) consist of an infinite number of points and some kinds of subjective judgments should be added to the quantitative analyses by the DM. Namely, the DM must select a satisficing solution from an M-Pareto optimal solution set based on a subjective value judgement.

10.1.3 Minimax problems

Having determined the membership functions for each of the objective functions, in order to generate a candidate for the satisficing solution which is also M-Pareto optimal, the DM is asked to specify the reference levels of achievement of the membership functions, called reference membership levels [78]. For the DM's reference membership levels $\bar{\mu}_i$, $i = 1, \ldots, k$, the corresponding M-Pareto optimal solution, which is, in the minimax sense, nearest to the requirement or better than that if the reference levels are attainable, is obtained by solving the following minimax problem:

$$\left.\begin{array}{ll} \text{minimize} & \max\limits_{i=1,\ldots,k} \{\bar{\mu}_i - \mu_i(z_i(\boldsymbol{x}, \boldsymbol{c}_i, \boldsymbol{d}_i))\} \\ \text{subject to} & \boldsymbol{x} \in X. \end{array}\right\} \tag{10.4}$$

Noting $q_i(\boldsymbol{x}, \boldsymbol{d}_i) > 0, i = 1, \ldots, k$, for $\boldsymbol{x} \in X$, this problem can be equivalently transformed as

$$\left.\begin{array}{ll} \text{minimize} & v \\ \text{subject to} & \bar{\mu}_i - \mu_i(z_i(\boldsymbol{x}, \boldsymbol{c}_i, \boldsymbol{d}_i)) \leq v, \;\; i = 1, \ldots, k \\ & \boldsymbol{x} \in X. \end{array}\right\} \tag{10.5}$$

For notational convenience, denote the strictly monotone decreasing function for the fuzzy min and the right function of the fuzzy equal by $D_{iR}(z_i(\boldsymbol{x}, \boldsymbol{c}_i, \boldsymbol{d}_i))$, $(i \in I_1 \cup I_{3R})$ and the strictly monotone increasing function for the fuzzy max and the left function of the fuzzy equal by $D_{iL}(z_i(\boldsymbol{x}, \boldsymbol{c}_i, \boldsymbol{d}_i))$, $(i \in I_2 \cup I_{3L})$. Then in order to solve the formulated problem on the basis of the linear programming method with decomposition procedures, we first convert each constraint $\bar{\mu}_i - \mu_i(z_i(\boldsymbol{x}, \boldsymbol{c}_i, \boldsymbol{d}_i)) \leq v$, $i = 1, \ldots, k$, of the minimax problem (10.5) into the following form using the strictly monotone decreasing or increasing property of $D_{iR}(\cdot)$ and $D_{iL}(\cdot)$ together with $q_i(\boldsymbol{x}, \boldsymbol{d}_i) > 0$, $i = 1, \ldots, k$.

$$\left.\begin{array}{l} p_i(\boldsymbol{x}, \boldsymbol{c}_i) \leq D_{iR}^{-1}(\bar{\mu}_i - v) q_i(\boldsymbol{x}, \boldsymbol{d}_i) \;\; i \in I_1 \cup I_3 \\ p_i(\boldsymbol{x}, \boldsymbol{c}_i) \geq D_{iL}^{-1}(\bar{\mu}_i - v) q_i(\boldsymbol{x}, \boldsymbol{d}_i) \;\; i \in I_2 \cup I_3 \end{array}\right\} \quad (10.6)$$

Then we can obtain an optimal solution of the problem (10.5) by solving the following linear fractional programming problem:

$$\left.\begin{array}{ll} \text{minimize} & v \\ \text{subject to} & p_i(\boldsymbol{x}, \boldsymbol{c}_i) \leq D_{iR}^{-1}(\bar{\mu}_i - v) q_i(\boldsymbol{x}, \boldsymbol{d}_i), \;\; i \in I_1 \cup I_3 \\ & p_i(\boldsymbol{x}, \boldsymbol{c}_i) \geq D_{iL}^{-1}(\bar{\mu}_i - v) q_i(\boldsymbol{x}, \boldsymbol{d}_i), \;\; i \in I_2 \cup I_3 \\ & \boldsymbol{x} \in X. \end{array}\right\} \quad (10.7)$$

It is important to note here that in this formulation, if the value of v is fixed, it can be reduced to a set of linear inequalities. Obtaining the optimal solution v^* to the above problem is equivalent to determining the minimum value of v so that there exists an admissible set satisfying the constraints of (10.7). Since v satisfies $\bar{\mu}_{\max} - 1 \leq v \leq \bar{\mu}_{\max}$, where $\bar{\mu}_{\max}$ denotes the maximum value of $\bar{\mu}_i$, $i =, \ldots, k$, we have the following method for solving this problem by combining use of the bisection method and phase one of the simplex method of linear programming.

Step 1: Set $v = \bar{\mu}_{\max}$ and test whether an admissible set satisfying the constraints of (10.7) exists or not using phase one of the simplex method. If an admissible set exists, proceed. Otherwise, the DM must reassess a membership function.

Step 2: Set $v = \bar{\mu}_{\max} - 1$ and test whether an admissible set satisfying the constraints of (10.7) exists or not using phase one of the simplex method. If an admissible set exists, set $v^* = \bar{\mu}_{\max} - 1$. Otherwise go to the next step, since the minimum v which satisfies the constraints of (10.7) exists between $\bar{\mu}_{\max} - 1$ and $\bar{\mu}_{\max}$.

Step 3: For the initial value of $v_1 = \bar{\mu}_{\max} - 0.5$, update the value of v using the bisection method as follows:

$v_{n+1} = v_n - 1/2^{n+1}$ if admissible set exists for v_n,

$v_{n+1} = v_n + 1/2^{n+1}$ if no admissible set exists for v_n. Namely, for each v, test whether an admissible set of (10.7) exists or not using phase one of the simplex method and determine the minimum value of v satisfying the constraints of (10.7).

It should be noted here that for the fixed value v, the problem (10.7) becomes a linear programming problem with the block angular structure as shown in Figure 10.1; the Dantzig-Wolfe decomposition method is applicable for phase one of the simplex method.

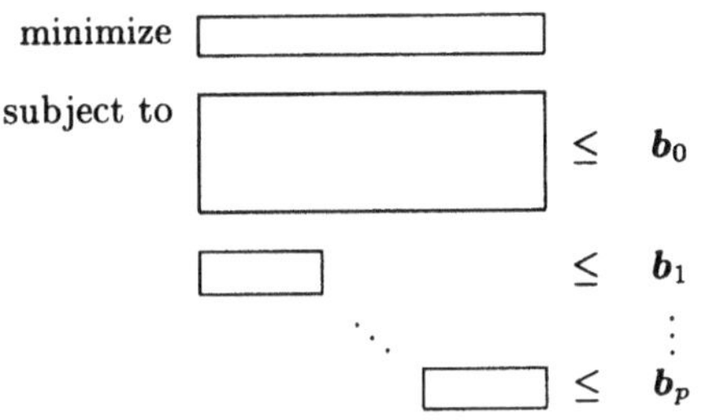

Figure 10.1. Block angular structure with coupling constraints.

In this way, we can determine the optimal solution v^*. Then, in order to determine the optimal value of $\boldsymbol{x}^*$ for the optimal solution v^*, the DM selects an appropriate standing objective from the objectives $z_i(\cdot,\cdot,\cdot), i = 1,\ldots,k$. For convenience, in the following, let it be $z_1(\cdot,\cdot,\cdot)$. Then the following linear fractional programming problem is solved for $v = v^*$:

$$\left.\begin{array}{ll} \text{minimize} & z_1(\boldsymbol{x}, \boldsymbol{c}_1, \boldsymbol{d}_1) \\ \text{subject to} & p_i(\boldsymbol{x}, \boldsymbol{c}_i) \le D_{iR}^{-1}(\bar{\mu}_i - v^*) q_i(\boldsymbol{x}, \boldsymbol{d}_i), \ i \in I_1 \cup I_3 (i \neq 1) \\ & p_i(\boldsymbol{x}, \boldsymbol{c}_i) \ge D_{iL}^{-1}(\bar{\mu}_i - v^*) q_i(\boldsymbol{x}, \boldsymbol{d}_i), \ i \in I_2 \cup I_3 \\ & \boldsymbol{x} \in X. \end{array}\right\} \tag{10.8}$$

In order to solve this linear fractional programming problem, if we use Charnes-Cooper's variable transformation [7]

$$t = \frac{1}{q_1(\boldsymbol{x}, \boldsymbol{d}_1)}, \quad \boldsymbol{y} = \begin{pmatrix} \boldsymbol{x} \\ 1 \end{pmatrix} t \tag{10.9}$$

the problem (10.8) can be equivalently transformed as

$$\left.\begin{array}{ll} \text{minimize} & \boldsymbol{c}_1\boldsymbol{y} \\ \text{subject to} & \boldsymbol{c}_i\boldsymbol{y} \leq D_{iR}^{-1}(\bar{\mu}_i - v^*)\boldsymbol{d}_i\boldsymbol{y}, \ i \in I_1 \cup I_3 (i \neq 1) \\ & \boldsymbol{c}_i\boldsymbol{y} \geq D_{iL}^{-1}(\bar{\mu}_i - v^*)\boldsymbol{d}_i\boldsymbol{y}, \ i \in I_2 \cup I_3 \\ & \boldsymbol{d}_1\boldsymbol{y} = 1 \\ & \boldsymbol{y} \in Y. \end{array}\right\} \qquad (10.10)$$

It is significant to note here that this problem involves not only coupling constraints but also coupling variables as shown in Figure 10.2; the Dantzig-Wolfe decomposition method is not applicable. Fortunately, however, instead of the Dantzig-Wolfe decomposition method, we can apply Ritter's partitioning procedure [72] for solving it.

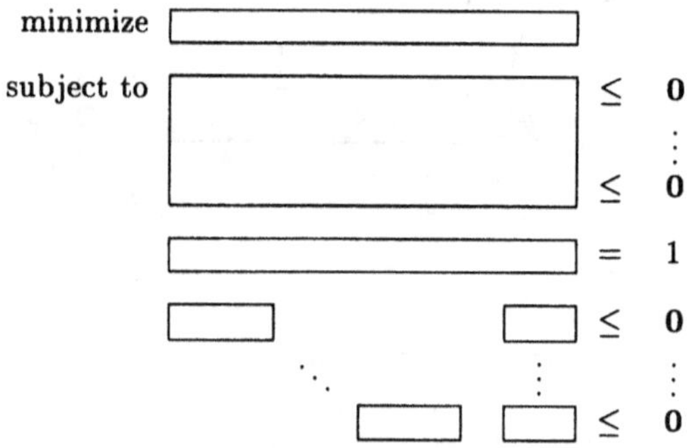

Figure 10.2. Block angular structure with coupling constraints and variables.

It should be noted here that for generating M-Pareto optimal solutions, uniqueness of the solution must be verified. In general, however, it is not easy to check numerically whether an optimal solution to (10.10) is unique. Consequently, in order to test the M-Pareto optimality of a current optimal solution $\boldsymbol{x}^*$, we formulate and solve the following linear programming problem:

$$\left.\begin{array}{ll} \text{maximize} & w = \displaystyle\sum_{i=1}^{k} \varepsilon_i \\ \text{subject to} & p_i(\boldsymbol{x}, \boldsymbol{c}_i) + \varepsilon_i = z_i(\boldsymbol{x}^*, \boldsymbol{c}_i, \boldsymbol{d}_i) q_i(\boldsymbol{x}, \boldsymbol{d}_i), \ i \in I_1 \cup I_{3R} \\ & p_i(\boldsymbol{x}, \boldsymbol{c}_i) - \varepsilon_i = z_i(\boldsymbol{x}^*, \boldsymbol{c}_i, \boldsymbol{d}_i) q_i(\boldsymbol{x}, \boldsymbol{d}_i), \ i \in I_2 \cup I_{3L} \\ & \boldsymbol{x} \in X, \ \varepsilon_i \geq 0, \ i = 1, \ldots, k. \end{array}\right\} \qquad (10.11)$$

This problem preserves the block angular structure involving only coupling constraints; the Dantzig-Wolfe decomposition method is applicable.

Let $\bar{\boldsymbol{x}}$ and $\bar{\varepsilon}_i$ be an optimal solution to (10.11). If $w = 0$, that is, $\varepsilon_i = 0$, then $\boldsymbol{x}^*$ is an M-Pareto optimal solution. In case of $\bar{w} > 0$ and consequently at least one $\bar{\varepsilon}_i > 0$, we perform the following operations.

Step 1: For an optimal solution to problem (10.11), solve the following problem for any j such that $\bar{\varepsilon}_j > 0$.

$$\left.\begin{array}{ll} \text{minimize} & z_j(\boldsymbol{x}, \boldsymbol{c}_j, \boldsymbol{d}_j) \\ \text{subject to} & z_i(\boldsymbol{x}, \boldsymbol{c}_i, \boldsymbol{d}_i) = z_i(\bar{\boldsymbol{x}}, \boldsymbol{c}_i, \boldsymbol{d}_i), \ \{i \in I_1 \cup I_{3R} \mid \bar{\varepsilon}_i = 0\} \\ & z_i(\boldsymbol{x}, \boldsymbol{c}_i, \boldsymbol{d}_i) = z_i(\bar{\boldsymbol{x}}, \boldsymbol{c}_i, \boldsymbol{d}_i), \ \{i \in I_2 \cup I_{3L} \mid \bar{\varepsilon}_i = 0\} \\ & z_i(\boldsymbol{x}, \boldsymbol{c}_i, \boldsymbol{d}_i) \leq z_i(\bar{\boldsymbol{x}}, \boldsymbol{c}_i, \boldsymbol{d}_i), \ \{i \in I_1 \cup I_{3R} \mid \bar{\varepsilon}_i > 0\} \\ & z_i(\boldsymbol{x}, \boldsymbol{c}_i, \boldsymbol{d}_i) \geq z_i(\bar{\boldsymbol{x}}, \boldsymbol{c}_i, \boldsymbol{d}_i), \ \{i \in I_2 \cup I_{3L} \mid \bar{\varepsilon}_i > 0\} \\ & \boldsymbol{x} \in X. \end{array}\right\} \tag{10.12}$$

This problem involves coupling constraints and variables after the Charnes-Cooper's variable transformation, we can apply Ritter's partitioning procedure [72] for solving it.

Step 2: Test the M-Pareto optimality for the solution to the problem (10.12) by solving the problem (10.11).

Step 3: If $w = 0$, stop. Otherwise, return to Step 1.

Repeating this process at least $k - 1$ iterations, an M-Pareto optimal solution can be obtained.

10.1.4 Interactive fuzzy multiobjective programming

Now given the M-Pareto optimal solution for the reference levels specified by the DM by solving the corresponding minimax problem, the DM must either be satisfied with the current M-Pareto optimal solution, or update reference membership levels. In order to help the DM express a degree of preference, trade-off information between a standing membership function and each of the other membership functions is very useful. Such trade-off information is easily obtainable since it is closely related to the simplex multipliers of the problem (10.10).

To derive the trade-off information, we define the following Lagrangian function L corresponding to the problem (10.10).

$$\begin{aligned} L = \boldsymbol{c}_1 \boldsymbol{y} + \eta(\boldsymbol{d}_i \boldsymbol{y} - 1) + \sum_{i \in I_1 \cup I_{3R}} \pi_{iR} \{\boldsymbol{c}_i \boldsymbol{y} - D_{iR}^{-1}(\bar{\mu}_i - v^*) \boldsymbol{d}_i \boldsymbol{y}\} \\ + \sum_{i \in I_2 \cup I_{3L}} \pi_{iL} \{D_{iL}^{-1}(\bar{\mu}_i - v^*) \boldsymbol{d}_i \boldsymbol{y} - \boldsymbol{c}_i \boldsymbol{y}\} \end{aligned}$$

$$+\sum_{j=1}^{m_0} \lambda_j(\boldsymbol{a}_{j1}, \ldots, \boldsymbol{a}_{jp}, -b_j)\boldsymbol{y} \tag{10.13}$$

where η, π_{iL}, π_{iR}, λ_j are simplex multipliers corresponding to the constraints in the problem (10.10). Then using the result in Haimes and Chankong [31], the following expression holds:

$$-\frac{\partial(z_1(\boldsymbol{x}, \boldsymbol{c}_1, \boldsymbol{d}_1))}{\partial(z_i(\boldsymbol{x}, \boldsymbol{c}_i, \boldsymbol{d}_i))} = \pi_{iR}\boldsymbol{d}_i\boldsymbol{y}, \quad i \in I_1 \cup I_{3R},\ i \neq 1, \tag{10.14}$$

$$-\frac{\partial(z_1(\boldsymbol{x}, \boldsymbol{c}_1, \boldsymbol{d}_1))}{\partial(z_i(\boldsymbol{x}, \boldsymbol{c}_i, \boldsymbol{d}_i))} = -\pi_{iL}\boldsymbol{d}_i\boldsymbol{y}, \quad i \in I_2 \cup I_{3L}. \tag{10.15}$$

Furthermore, using the strictly monotone decreasing or increasing property of $D_{iR}(\cdot)$ and $D_{iL}(\cdot)$ together with the chain rule, if $D_{iR}(\cdot)$ and $D_{iL}(\cdot)$ are differentiable at the optimal solution to (10.10), it holds that

$$-\frac{\partial(\mu_1(\boldsymbol{x}, \boldsymbol{c}_1, \boldsymbol{d}_1))}{\partial(\mu_i(\boldsymbol{x}, \boldsymbol{c}_i, \boldsymbol{d}_i))} = \pi_{iR}\boldsymbol{d}_i\boldsymbol{y}\frac{D'_{1R}(z_1(\boldsymbol{x}, \boldsymbol{c}_1, \boldsymbol{d}_1))}{D'_{iR}(z_i(\boldsymbol{x}, \boldsymbol{c}_i, \boldsymbol{d}_i))}, \quad i \in I_1 \cup I_{3R},\ i \neq 1, \tag{10.16}$$

$$-\frac{\partial(\mu_1(\boldsymbol{x}, \boldsymbol{c}_1, \boldsymbol{d}_1))}{\partial(\mu_i(\boldsymbol{x}, \boldsymbol{c}_i, \boldsymbol{d}_i))} = -\pi_{iL}\boldsymbol{d}_i\boldsymbol{y}\frac{D'_{1R}\left(z_1(\boldsymbol{x}, \boldsymbol{c}^L_{1\alpha}, \boldsymbol{d}^R_{1\alpha})\right)}{D'_{iL}(z_i(\boldsymbol{x}, \boldsymbol{c}_i, \boldsymbol{d}_i))}, \quad i \in I_1 \cup I_{3R}. \tag{10.17}$$

It should be noted here that in order to obtain the trade-off rate information from (10.16) and (10.17), all constraints of the problem (10.10) must be active for the current optimal solution $\boldsymbol{x}^*$, and $\boldsymbol{x}^*$ must satisfy the M-Pareto optimality test. Therefore, if there are inactive constraints, it is necessary to replace $\bar{\mu}_i$ for inactive constraints by $D_{iR}(\boldsymbol{c}_i\boldsymbol{y}^*/\boldsymbol{d}_i\boldsymbol{y}^*)+v^*$, $i \in I_1 \cup I_{3R} (i \neq 1)$ or $D_{iL}(\boldsymbol{c}_i\boldsymbol{y}^*/\boldsymbol{d}_i\boldsymbol{y}^*) + v^*$, $i \in I_2 \cup I_{3L}$ and solve the corresponding problem (10.10) for obtaining the simplex multipliers.

Following the preceding discussions, we can now construct the interactive algorithm in order to derive the satisficing solution for the DM from the M-Pareto optimal solution set. The steps marked with an asterisk involve interaction with the DM.

Interactive fuzzy multiobjective linear fractional programming

Step 0: Calculate the individual minimum and maximum of each objective function under the given constraints using Ritter's partitioning procedure.

Step 1*: Elicit a membership function $\mu_i(z_i(\boldsymbol{x}, \boldsymbol{c}_i, \boldsymbol{d}_i))$ from the DM for each of the objective functions.

Step 2: Set the initial reference membership levels $\bar{\mu}_i = 1$, $i = 1, \ldots, k$.

Step 3: For the reference membership levels specified by the DM, solve the minimax problem and perform the M-Pareto optimality test through the Dantzig-Wolfe decomposition method and Ritter's partitioning procedure.

Step 4*: The DM is supplied with the corresponding M-Pareto optimal solution and the trade-off rates between the membership functions. If the DM is satisfied with the current membership function values of the M-Pareto optimal solution, stop. Otherwise, the DM must update the reference levels by considering the current values of the membership functions together with the trade-off rates between the membership functions and return to Step 3.

Here it should be stressed to the DM that any improvement of one membership function can be achieved only at the expense of at least one of the other membership functions.

10.1.5 Numerical example

To demonstrate the feasibility of the proposed interactive decision making method, consider a three-objective linear fractional programming problem with a block angular structure as a numerical example. To be more specific, this example involves 8 variables, 13 coupling constraints, 8 block constraints (2 constraints for each 4 blocks) and is formulated as

$$
\begin{array}{rlllll}
\text{fuzzy min} & z_1(\boldsymbol{x}, \boldsymbol{c}_1, \boldsymbol{d}_1) & & & & \\
\text{fuzzy max} & z_2(\boldsymbol{x}, \boldsymbol{c}_2, \boldsymbol{d}_2) & & & & \\
\text{fuzzy equal} & z_3(\boldsymbol{x}, \boldsymbol{c}_3, \boldsymbol{d}_3) & & & & \\
\text{subject to} & A_1\boldsymbol{x}_1 & +A_2\boldsymbol{x}_2 & +A_3\boldsymbol{x}_3 & +A_4\boldsymbol{x}_4 & [0, \leq \text{ or } \geq]\ \boldsymbol{b}_0 \\
& B_1\boldsymbol{x}_1 & & & & [1, \leq \text{ or } \geq]\ \boldsymbol{b}_1 \\
& & B_2\boldsymbol{x}_2 & & & [2, \leq \text{ or } \geq]\ \boldsymbol{b}_2 \\
& & & B_3\boldsymbol{x}_3 & & [3, \leq \text{ or } \geq]\ \boldsymbol{b}_3 \\
& & & & B_4\boldsymbol{x}_4 & [4, \leq \text{ or } \geq]\ \boldsymbol{b}_4 \\
& & \boldsymbol{x}_i \geq \boldsymbol{0}, \ i = 1, \ldots, 4, & & &
\end{array}
$$

where

$$
z_i(\boldsymbol{x}, \boldsymbol{c}_i, \boldsymbol{d}_i) = \frac{\boldsymbol{c}_{i1}\boldsymbol{x}_1 + \boldsymbol{c}_{i2}\boldsymbol{x}_2 + \boldsymbol{c}_{i3}\boldsymbol{x}_3 + \boldsymbol{c}_{i4}\boldsymbol{x}_4 + c_{i5}}{\boldsymbol{d}_{i1}\boldsymbol{x}_1 + \boldsymbol{d}_{i2}\boldsymbol{x}_2 + \boldsymbol{d}_{i3}\boldsymbol{x}_3 + \boldsymbol{d}_{i4}\boldsymbol{x}_4 + d_{i5}}, \quad i = 1, 2, 3,
$$

and

$$\boldsymbol{x}_1 = (x_1, x_2), \boldsymbol{x}_2 = (x_3, x_4), \boldsymbol{x}_3 = (x_5, x_6), \boldsymbol{x}_4 = (x_7, x_8),$$
$$\boldsymbol{c}_{11} = (-0.3, -0.3), \boldsymbol{c}_{12} = (1.8, 2.0), \boldsymbol{c}_{13} = (2.9, 1.0), \boldsymbol{c}_{14} = (6.0, 0.8),$$
$$\boldsymbol{c}_{21} = (-0.4, 0.7), \boldsymbol{c}_{22} = (2.0, -0.4), \boldsymbol{c}_{23} = (2.5, 6.5), \boldsymbol{c}_{24} = (1.0, 0.2),$$
$$\boldsymbol{c}_{31} = (0.3, 1.5), \boldsymbol{c}_{32} = (0.9, 2.4), \boldsymbol{c}_{33} = (3.5, 0.7), \boldsymbol{c}_{34} = (4.4, 0.5)$$
$$\boldsymbol{d}_{11} = (2.0, 0.2), \boldsymbol{d}_{12} = (-0.3, 0.5), \boldsymbol{d}_{13} = (-1.9, 0.7), \boldsymbol{d}_{14} = (3.4, -0.4),$$
$$\boldsymbol{d}_{21} = (1.0, 1.3), \boldsymbol{d}_{22} = (2.4, 0.5), \boldsymbol{d}_{23} = (0.7, 4.1), \boldsymbol{d}_{24} = (0.2, 1.9),$$
$$\boldsymbol{d}_{31} = (1.5, 2.3), \boldsymbol{d}_{32} = (0.2, 0.1), \boldsymbol{d}_{33} = (1.1, 0.8), \boldsymbol{d}_{34} = (1.2, 0.6)$$
$$c_{15} = 10.0, c_{25} = 0.8, c_{35} = 3.6, d_{15} = 3.4, d_{25} = 0.5,\ d_{35} = 0.8,$$

$$A_1 = \begin{bmatrix} 1.0 & 0.0 \\ 0.0 & 1.0 \\ 1.0 & 1.0 \\ 1.0 & 1.0 \\ 0.0 & 0.0 \\ 0.0 & 0.0 \\ 2.0 & 2.0 \\ 1.0 & 1.0 \\ 0.0 & 0.0 \\ 0.0 & 0.0 \\ 3.5 & -1.5 \\ -1.0 & 4.2 \\ 1.2 & -0.4 \end{bmatrix}, A_2 = \begin{bmatrix} 1.0 & 0.0 \\ 0.0 & 1.0 \\ 0.0 & 0.0 \\ 0.0 & 0.0 \\ 1.0 & 1.0 \\ 1.0 & 1.0 \\ 1.0 & 1.0 \\ 6.0 & 6.0 \\ 0.0 & 0.0 \\ 0.0 & 0.0 \\ 1.0 & 2.3 \\ 5.0 & -1.4 \\ 0.7 & 8.0 \end{bmatrix}, A_3 = \begin{bmatrix} 1.0 & 0.0 \\ 0.0 & 1.0 \\ 1.0 & 1.0 \\ 1.0 & 1.0 \\ 0.0 & 0.0 \\ 0.0 & 0.0 \\ 0.0 & 0.0 \\ 0.0 & 0.0 \\ 3.0 & 3.0 \\ 1.0 & 1.0 \\ -1.7 & 4.0 \\ -3.1 & 1.0 \\ -2.1 & -0.8 \end{bmatrix}, A_4 = \begin{bmatrix} 1.0 & 0.0 \\ 0.0 & 1.0 \\ 0.0 & 0.0 \\ 0.0 & 0.0 \\ 1.0 & 1.0 \\ 1.0 & 1.0 \\ 0.0 & 0.0 \\ 0.0 & 0.0 \\ 1.0 & 1.0 \\ 2.0 & 2.0 \\ 3.5 & -1.2 \\ -1.3 & 1.0 \\ 1.0 & 2.5 \end{bmatrix}$$

$$B_1 = \begin{bmatrix} 1.0 & 1.0 \\ -1.0 & 1.0 \end{bmatrix}, \quad B_2 = \begin{bmatrix} 1.0 & 1.0 \\ 1.0 & -1.0 \end{bmatrix}, \quad B_3 = \begin{bmatrix} 1.0 & 1.0 \\ 1.0 & -1.0 \end{bmatrix}, \quad B_4 = \begin{bmatrix} 1.0 & 1.0 \\ 1.0 & -1.0 \end{bmatrix}$$

$$[0, \leq \text{or} \geq] = \begin{bmatrix} \leq \\ \leq \\ \leq \\ \geq \\ \leq \\ \geq \\ \leq \\ \leq \\ \leq \\ \leq \\ \geq \\ \leq \\ \leq \end{bmatrix}, \quad \boldsymbol{b}_0 = \begin{bmatrix} 1500.0 \\ 1500.0 \\ 1500.0 \\ 500.0 \\ 1000.0 \\ 300.0 \\ 3000.0 \\ 1000.0 \\ 800.0 \\ 1400.0 \\ 800.0 \\ 200.0 \\ 800.0 \end{bmatrix}$$

$$[1, \leq \text{or} \geq] = \begin{bmatrix} \leq \\ \leq \end{bmatrix}, \quad \boldsymbol{b}_1 = \begin{bmatrix} 1000.0 \\ 300.0 \end{bmatrix}, \quad [2, \leq \text{or} \geq] = \begin{bmatrix} \leq \\ \leq \end{bmatrix}, \quad \boldsymbol{b}_2 = \begin{bmatrix} 200.0 \\ 100.0 \end{bmatrix}$$

$$[3, \leq \text{or} \geq] = \begin{bmatrix} \leq \\ \leq \end{bmatrix}, \quad \boldsymbol{b}_3 = \begin{bmatrix} 250.0 \\ 200.0 \end{bmatrix}, \quad [4, \leq \text{or} \geq] = \begin{bmatrix} \leq \\ \leq \end{bmatrix}, \quad \boldsymbol{b}_4 = \begin{bmatrix} 500.0 \\ 200.0 \end{bmatrix}$$

First calculating the individual minimum and maximum of each objective functions via Ritter's partitioning procedure yields the results shown in Table 10.1.

Table 10.1. Individual minimum and maximum

	minimum	maximum
$z_1(\boldsymbol{x})$	0.7080	4.1399
$z_2(\boldsymbol{x})$	0.2457	1.2946
$z_3(\boldsymbol{x})$	0.6934	2.0486

Now, for illustrative purposes, by considering the individual minimum and maximum, assume that the interaction with the hypothetical DM establishes the following membership functions and corresponding assessment values:

$$z_1 : \text{linear } (z_1^0, z_1^1) = (1.0, 0.8)$$
$$z_2 : \text{exponential } (z_2^0, z_2^{0.5}, z_2^1) = (0.3, 0.8, 1.2)$$
$$z_3 : \begin{cases} \text{left: exponential } (z_{3L}^0, z_{3L}^{0.5}, z_{3L}^1) = (0.7, 0.95, 1.0) \\ \text{right: exponential } (z_{3R}^0, z_{3R}^{0.5}, z_{3R}^1) = (0.7, 0.95, 1.0) \end{cases}$$

where, within the interval $[z_i^0, z_i^1]$, the linear and exponential membership functions are defined as [78]

$$\mu_i(z_i(\boldsymbol{x})) = \frac{z_i(\boldsymbol{x}) - z_i^0}{z_i^1 - z_i^0}$$

$$\mu_i(z_i(\boldsymbol{x})) = a_i \left[1 - \exp\left\{-\alpha_i \frac{z_i(\boldsymbol{x}) - z_i^0}{z_i^1 - z_i^0}\right\}\right],$$

where $a_i > 1, \alpha_i > 0$ or $a_i < 0, \alpha_i < 0$.

The minimax problem is solved for the initial membership values 1 using the Dantzig-Wolfe decomposition method and Ritter's partitioning procedure, and the DM is supplied with the corresponding Pareto optimal solution and the trade-off rates between the membership functions as shown in the second column of Table 10.2.

On the basis of such information, suppose that the DM updates the reference levels to $(1.0, 0.9, 1.0)$, improving the satisfaction levels for μ_1 and μ_3 at the expense of μ_2. For the updated reference level, the corresponding results are shown in the third column of Table 10.2.

The same procedures continues in this fashion and in this example, at the fourth iteration, the satisficing solution for the DM is derived as shown in

Table 10.2. Interactive processes for example

iteration	1	2	3	4
$\bar{\mu}_1$	1.0	1.0	1.0	1.0
$\bar{\mu}_2$	1.0	0.9	0.9	0.8
$\bar{\mu}_3$	1.0	1.0	0.95	0.9
z_1	0.8648	0.8478	0.8473	0.8303
z_2	0.9500	0.9377	0.9399	0.9273
z_3	1.0565	1.0394	1.0487	1.0325
μ_1	0.6760	0.7610	0.7637	0.8485
μ_2	0.6760	0.6610	0.6637	0.6485
μ_3	0.6760	0.7610	0.7137	0.7985
$-\partial\mu_1/\partial\mu_2$	9.4140	8.8173	8.6852	8.1261
$-\partial\mu_1/\partial\mu_3$	0.6977	0.5350	0.5760	0.4393

the fifth column of Table 10.2, which becomes

$$\begin{aligned}
&x_1 = 427.39, \quad x_2 = 0.0, \quad x_3 = 95.43, \quad x_4 = 0.0,\\
&x_5 = 0.0, \quad x_6 = 198.48, \quad x_7 = 156.74, \quad x_8 = 47.83,\\
&z_1(\boldsymbol{x}) = 0.8303, \quad z_2(\boldsymbol{x}) = 0.9273, \quad z_3(\boldsymbol{x}) = 1.0325,\\
&\mu_1(z_1(\boldsymbol{x})) = 0.8485, \quad \mu_2(z_2(\boldsymbol{x})) = 0.6485, \quad \mu_3(z_3(\boldsymbol{x})) = 0.7985.
\end{aligned}$$

10.1.6 Conclusion

In this section, we focused on multiobjective linear fractional programming problems with the block angular structure and proposed an interactive decision making method for deriving the satisficing solution for the decision maker (DM) from the (M-) Pareto optimal solution set. In our interactive decision making method, after determining the membership functions, if the DM specifies the reference membership levels, the corresponding (M-) Pareto optimal solution can be easily obtained by solving the minimax problems by applying the Dantzig-Wolfe decomposition method and Ritter's partitioning procedure. In this way, the satisficing solution for the DM can be derived efficiently from (M-) Pareto optimal solutions by updating the reference membership levels based on the current values of the (M-) Pareto optimal solution together with the trade-off information between the membership functions. An illustrative numerical example demonstrated the feasibility of the proposed method.

It is significant to point out here that all the results presented in this section have already been extended by the authors to deal with large scale multiobjective linear fractional programming problems with block angular structures involving fuzzy parameters. A successful generalization along this line can be found in Sakawa and Kato [88, 89, 90], and the interested readers might refer to them for details.

10.2 Large scale fuzzy multiobjective nonlinear programming

As a nonlinear generalization of Chapter 5, in this section, we formulate large scale multiobjective nonlinear programming problems with the block angular structure by incorporating the fuzzy goals of the decision maker and propose both the dual decomposition method [121] and primal decomposition method [122] for deriving the satisficing solution for the decision maker (DM).

In the dual decomposition method, having determined the corresponding membership functions, following the fuzzy decision of Bellman and Zadeh for combining them, an augmented primal problem and the corresponding augmented dual problem are formulated. Under some suitable assumptions, it is shown that the augmented dual problem is decomposable with respect to the decision variables and hence it can be reduced to one master problem and a number of independent subproblems. As a result a two-level optimization algorithm to solve the augmented dual problem is developed for deriving the satisficing solution for the DM. An illustrative numerical example demonstrated the feasibility of the proposed method.

Unfortunately, however, in the dual decomposition method, the coupling constraints are not satisfied at any cycle until a convergence is achieved. For this reason, this approach is sometimes called the nonfeasible decomposition. To decrease such drawbacks of a dual decomposition method, we also present a primal decomposition approach by right-hand-side allocation for multiobjective nonlinear programming problems with block angular structure by incorporating the fuzzy goals of the DM . In the primal decomposition method, having elicited the corresponding linear membership functions through the interaction with the DM, if we adopt the add-operator for combining them, it is shown that the the original problem can be decomposed into several subproblems by introducing the coupling variables, which can be interpreted as right-hand-side allocations. Then a two-level optimization algorithm for deriving the satisficing solution for the DM is proposed. An illustrative numerical example is provided to indicate the feasibility of the proposed method.

10.2.1 Multiobjective nonlinear programming problems with block angular structures

As a nonlinear extension of Chapter 5, consider a large scale multiobjective nonlinear programming problem with the following block angular structure:

$$\left.\begin{array}{rrcl}
\text{minimize} & f_1(\boldsymbol{x}) & = & f_{11}(\boldsymbol{x}_1) + f_{12}(\boldsymbol{x}_2) + \cdots + f_{1p}(\boldsymbol{x}_p) \\
\text{minimize} & f_2(\boldsymbol{x}) & = & f_{21}(\boldsymbol{x}_1) + f_{22}(\boldsymbol{x}_2) + \cdots + f_{2p}(\boldsymbol{x}_p) \\
 & \vdots & & \vdots \\
\text{minimize} & f_k(\boldsymbol{x}) & = & f_{k1}(\boldsymbol{x}_1) + f_{k2}(\boldsymbol{x}_2) + \cdots + f_{kp}(\boldsymbol{x}_p) \\
\text{subject to} & g_1(\boldsymbol{x}) & = & g_{11}(\boldsymbol{x}_1) + g_{12}(\boldsymbol{x}_2) + \cdots + g_{1p}(\boldsymbol{x}_p) \leq 0 \\
 & g_2(\boldsymbol{x}) & = & g_{21}(\boldsymbol{x}_1) + g_{22}(\boldsymbol{x}_2) + \cdots + g_{2p}(\boldsymbol{x}_p) \leq 0 \\
 & \vdots & & \vdots \\
 & g_{m_0}(\boldsymbol{x}) & = & g_{m_01}(\boldsymbol{x}_1) + g_{m_02}(\boldsymbol{x}_2) + \cdots + g_{m_0p}(\boldsymbol{x}_p) \leq 0 \\
 & & & \boldsymbol{h}_1(\boldsymbol{x}_1) \leq \boldsymbol{0} \\
 & & & \qquad \boldsymbol{h}_2(\boldsymbol{x}_2) \leq \boldsymbol{0} \\
 & & & \qquad\qquad \ddots \qquad \vdots \\
 & & & \qquad\qquad\qquad \boldsymbol{h}_p(\boldsymbol{x}_p) \leq \boldsymbol{0}
\end{array}\right\} \quad (10.18)$$

where $f_i(\boldsymbol{x})$, $i = 1, \ldots, k$, are k distinct objective functions of the decision vector $\boldsymbol{x} = (\boldsymbol{x}_1, \ldots, \boldsymbol{x}_p) \in R^n$, $\boldsymbol{x}_j \in R^{n_j}$, $j = 1, \ldots, p$, are n_j dimensional vectors of decision variables, $g_\ell(\boldsymbol{x}) \leq \boldsymbol{0}$, $\ell = 1,, \ldots, m_0$, are m_0 coupling constraints, $\boldsymbol{h}_j(\boldsymbol{x}_j)$, $j = 1, \ldots, p$, are r_j dimensional constraints with respect to $\boldsymbol{x}_j$, and $f_{ij}(\boldsymbol{x}_j)$, $g_{\ell j}(\boldsymbol{x}_j)$, $i = 1, \ldots, k$, $\ell = 1, \ldots, m_0$, $j = 1, \ldots, p$, are nonlinear functions with respect to $\boldsymbol{x}_j$.

10.2.2 Dual decomposition method with fuzzy goals

In this subsection, for simplicity, we make the following convexity assumptions.

Assumption 10.2.1. The subset

$$S_j = \{\boldsymbol{x}_j \in R^{n_j} \mid \boldsymbol{h}_j(\boldsymbol{x}_j) \leq \boldsymbol{0}\} \neq \phi, \; j = 1, \ldots, p \quad (10.19)$$

is compact and convex

For notational convenience, define the set

$$S = \prod_{j=1,\ldots,p} S_j \subset R^n \tag{10.20}$$

where the symbol $\prod$ means the direct product. Observe that the set S becomes compact and convex.

Assumption 10.2.2. The objective functions $f_i(\boldsymbol{x})$, $i = 1, \ldots, k$, and the coupling constraints $g_\ell(\boldsymbol{x})$, $\ell = 1, \ldots, m_0$, are continuous and convex on the subset S_j.

Considering the vague nature of human's subjective judgments, it seems to be quite natural to assume that the decision maker (DM) may have a fuzzy goal for each of the objective functions [78, 142]. These fuzzy goals can be quantified by eliciting the corresponding membership functions $\mu_i(f_i(\boldsymbol{x})), i = 1, \ldots, k$ through the interaction with the DM. Concerning the membership functions $\mu_i(f_i(\boldsymbol{x}))$, $i = 1, \ldots, k$, we make the following assumption.

Assumption 10.2.3. The membership function $\mu_i(\cdot)$ for each of the objective functions is a surjection and strictly monotone decreasing continuous concave function on the subset S_j.

After determining the membership functions, if we assume the fuzzy decision of Bellman and Zadeh [3] is the proper representation of the DM's fuzzy preference, the corresponding satisficing solution for the DM can be derived by solving the following large scale fuzzy decision making problem:

$$\left.\begin{aligned} &\underset{\boldsymbol{x}=(x_1,\ldots,x_p)}{\text{maximize}} \left\{ \min_{i=1,\ldots,k} \mu_i(f_i(\boldsymbol{x})) \right\} \\ &\text{subject to } g_\ell(\boldsymbol{x}) = \sum_{\ell=1}^{p} g_{\ell j}(\boldsymbol{x}_j) \leq \mathbf{0}, \;\; \ell = 1, \ldots, m_0, \\ &\qquad \boldsymbol{x}_j \in S_j, \; j = 1, \ldots, p, \end{aligned}\right\} \tag{10.21}$$

or, equivalently, the following problem.

Primal problem (P)

$$\left.\begin{aligned} &\underset{(\boldsymbol{x},x_{p+1})=(x_1,\ldots,x_p,x_{p+1})}{\text{minimize}} \quad -x_{p+1} \\ &\text{subject to } f_i(\boldsymbol{x}) - {\mu_i}^{-1}(x_{p+1}) \;\leq 0, \;\; i = 1, \ldots, k, \\ &\qquad g_\ell(\boldsymbol{x}) = \sum_{j=1}^{p} g_{\ell j}(\boldsymbol{x}_j) \leq \mathbf{0}, \; \ell = 1, \ldots, m_0, \\ &\qquad \boldsymbol{x}_j \in S_j, \;\; j = 1, \ldots, p, \;\; x_{p+1} \in [0, 1], \end{aligned}\right\} \tag{10.22}$$

where $x_{p+1} \in R^1$ is an auxiliary variable and $\mu_i{}^{-1}(\cdot)$ is the inverse function of $\mu_i(\cdot)$. It is clear that such an inverse function always exists under Assumption 10.2.3. In the following, we call this formulation the primal problem (P).

When we deal with multiobjective nonlinear programming problems involving very large number of decision variables and/or constraints, the larger the overall problem becomes, the more difficult it becomes to solve the primal problem directly due to the high dimensionality of the problem. In order to overcome the high dimensionality of the primal problem (P), we introduce the dual decomposition method for large scale nonlinear programming problems proposed by Lasdon [57, 58].

Unfortunately, however, since the objective function in the primal problem (P) is not strictly convex, we cannot directly apply the dual decomposition method [57, 58] to the primal problem (P). In order to circumvent this difficulty, instead of dealing with the primal problem (P), we introduce the following augmented problem by adding the term ρx_{p+1}^2 to the objective function, where $\rho > 0$ is a sufficiently small positive constant.
Augmented primal problem (AP)

$$\left.\begin{array}{ll} \underset{(\boldsymbol{x}, x_{p+1})=(\boldsymbol{x}_1,\ldots,\boldsymbol{x}_p, x_{p+1})}{\text{minimize}} & -x_{p+1} + \rho x_{p+1}^2 \\ \text{subject to} & \displaystyle\sum_{j=1}^{p} f_{ij}(\boldsymbol{x}_j) - \mu_i{}^{-1}(x_{p+1}) \leq 0, \quad i = 1,\ldots,k \\ & g_\ell(\boldsymbol{x}) = \displaystyle\sum_{j=1}^{p} g_{\ell j}(\boldsymbol{x}_j) \leq 0, \ \ell = 1,\ldots,m_0, \\ & \boldsymbol{x}_j \in S_j, \ \ j = 1,\ldots,p \\ & x_{p+1} \in [0,1] \end{array}\right\} \tag{10.23}$$

It should be noted here that, for a sufficiently small positive $\rho > 0$, the optimal solution of the augmented primal problem (AP) approximately coincide with the optimal one of the primal problem (P).

For the augmented primal problem (AP), the corresponding Lagrangian function can be defined as:

$$\begin{aligned} L'(\boldsymbol{x}, x_{p+1}, \boldsymbol{\lambda}, \boldsymbol{\pi}) = & \sum_{j=1}^{p}\left\{\sum_{i=1}^{k} \lambda_i f_{ij}(\boldsymbol{x}_j) + \sum_{j=1}^{m_0} \pi_j g_{jj}(\boldsymbol{x}_j)\right\} \\ & + \left\{-x_{p+1} + \rho x_{p+1}^2 - \sum_{i=1}^{k} \lambda_i \mu_i{}^{-1}(x_{p+1})\right\} \end{aligned} \tag{10.24}$$

where $\boldsymbol{\lambda} = (\lambda_1, \ldots, \lambda_k)$ and $\boldsymbol{\pi} = (\pi_1, \ldots, \pi_{m_0})$ are vectors of Lagrange multipliers corresponding to the first and second constraints of the (AP), respectively.

Now we introduce the following sub-Lagrangian functions:

$$L_j(\boldsymbol{x}_j, \boldsymbol{\lambda}, \boldsymbol{\pi}) \triangleq \sum_{i=1}^{k} \lambda_i f_{ij}(\boldsymbol{x}_j) + \sum_{\ell=1}^{m_0} \pi_\ell g_{\ell j}(\boldsymbol{x}_j), \; j = 1, \ldots, p \quad (10.25)$$

$$L_{p+1}(x_{p+1}, \boldsymbol{\lambda}) \triangleq -x_{p+1} + \rho x_{p+1}^2 - \sum_{i=1}^{k} \lambda_i {\mu_i}^{-1}(x_{p+1}) \quad (10.26)$$

Then the Lagrangian function can be decomposed with respect to the decision variables $(\boldsymbol{x}_1, \ldots, \boldsymbol{x}_p, x_{p+1})$ as follows:

$$L(\boldsymbol{x}, x_{p+1}, \boldsymbol{\lambda}, \boldsymbol{\pi}) = \sum_{j=1}^{p} L_j(\boldsymbol{x}_j, \boldsymbol{\lambda}, \boldsymbol{\pi}) + L_{p+1}(x_{p+1}, \boldsymbol{\lambda}). \quad (10.27)$$

Using the sub-Lagrangian functions, the corresponding augmented dual problem (AD) can be formulated as follows:
Augmented dual problem (AD)

$$\underset{(\boldsymbol{\lambda}, \boldsymbol{\pi}) \in U}{\text{maximize}} \quad w(\boldsymbol{\lambda}, \boldsymbol{\pi}) \quad (10.28)$$

where the dual function $w(\boldsymbol{\lambda}, \boldsymbol{\pi})$ and the domain U of the Lagrange multipliers $(\boldsymbol{\lambda}, \boldsymbol{\pi})$ are defined as

$$\begin{aligned} w(\boldsymbol{\lambda}, \boldsymbol{\pi}) \;\triangleq\; & \underset{(\boldsymbol{x}, x_{p+1})}{\text{minimize}} \;\; \sum_{j=1}^{p} L_j(\boldsymbol{x}_j, \boldsymbol{\lambda}, \boldsymbol{\pi}) + L_{p+1}(x_{p+1}, \boldsymbol{\lambda}) \\ & \text{subject to} \;\; \boldsymbol{x}_j \in S_j, \; j = 1, \ldots, p \\ & \qquad\qquad\quad x_{p+1} \in [0, 1] \end{aligned} \quad (10.29)$$

$$U \triangleq \left\{ (\boldsymbol{\lambda}, \boldsymbol{\pi}) \in R^{k+m_0} \;\middle|\; \boldsymbol{\lambda} \geq \boldsymbol{0}, \boldsymbol{\pi} \geq \boldsymbol{0}, \text{ the dual function } w(\boldsymbol{\lambda}, \boldsymbol{\pi}) \text{ exists} \right\} \quad (10.30)$$

Under Assumptions 10.2.1, 10.2.2 and 10.2.3, the relationships between the augmented primal problem (AP) and the augmented dual problem (AD) can be expressed by the following theorems [57, 58].

Theorem 10.2.1. *The domain of the dual function $w(\boldsymbol{\lambda}, \boldsymbol{\pi})$ is given by*

$$U = \left\{ (\boldsymbol{\lambda}, \boldsymbol{\pi}) \in R^{k+m_0} \;\middle|\; \boldsymbol{\lambda} \geq \boldsymbol{0}, \; \boldsymbol{\pi} \geq \boldsymbol{0} \right\}, \quad (10.31)$$

and the dual function $w(\boldsymbol{\lambda}, \boldsymbol{\pi})$ is concave.

Proof. From Assumptions 10.2.1, 10.2.2 and 10.2.3, the Lagrangian function $L(\boldsymbol{x}, x_{p+1}, \boldsymbol{\lambda}, \boldsymbol{\pi})$ is continuous with respect to $\boldsymbol{x} \in S$, $x_{p+1} \in [0,1]$ for all $(\boldsymbol{\lambda}, \boldsymbol{\pi}) \geq \mathbf{0}$, and S is closed and bounded. Then, by the Weierstrass theorem, we have $U = \left\{(\boldsymbol{\lambda}, \boldsymbol{\pi}) \in R^{k+m_0} \mid \boldsymbol{\lambda} \geq \mathbf{0},\ \boldsymbol{\pi} \geq \mathbf{0}\right\}$. Let $(\boldsymbol{\lambda}_1, \boldsymbol{\pi}_1) \in U$, $(\boldsymbol{\lambda}_2, \boldsymbol{\pi}_2) \in U$ and $0 \leq \alpha \leq 1$. Then $(\alpha\boldsymbol{\lambda}_1 + (1-\alpha)\boldsymbol{\lambda}_2, \alpha\boldsymbol{\pi}_1 + (1-\alpha)\boldsymbol{\pi}_2) \in U$ and

$$
\begin{aligned}
& w(\alpha\boldsymbol{\lambda}_1 + (1-\alpha)\boldsymbol{\lambda}_2, \alpha\boldsymbol{\pi}_1 + (1-\alpha)\boldsymbol{\pi}_2) \\
&= \min_{\boldsymbol{x} \in S, x_{p+1} \in [0,1]} L(\boldsymbol{x}, x_{p+1}, \alpha\boldsymbol{\lambda}_1 + (1-\alpha)\boldsymbol{\lambda}_2, \alpha\boldsymbol{\pi}_1 + (1-\alpha)\boldsymbol{\pi}_2) \\
&= \min_{\boldsymbol{x} \in S, x_{p+1} \in [0,1]} \left\{\alpha L(\boldsymbol{x}, x_{p+1}, \boldsymbol{\lambda}_1, \boldsymbol{\pi}_1) + (1-\alpha) L(\boldsymbol{x}, x_{p+1}, \boldsymbol{\lambda}_2, \boldsymbol{\pi}_2)\right\} \\
&\geq \alpha \min_{\boldsymbol{x} \in S, x_{p+1} \in [0,1]} L(\boldsymbol{x}, x_{p+1}, \boldsymbol{\lambda}_1, \boldsymbol{\pi}_1) + (1-\alpha) \min_{\boldsymbol{x} \in S, x_{p+1} \in [0,1]} L(\boldsymbol{x}, x_{p+1}, \boldsymbol{\lambda}_2, \boldsymbol{\pi}_2) \\
&= \alpha w(\boldsymbol{\lambda}_1, \boldsymbol{\pi}_1) + (1-\alpha) w(\boldsymbol{\lambda}_2, \boldsymbol{\pi}_2)
\end{aligned}
$$

This means that $w(\boldsymbol{\lambda}, \boldsymbol{\pi})$ is concave with respect to $(\boldsymbol{\lambda}, \boldsymbol{\pi}) \in U$.

Theorem 10.2.2. *The dual function $w(\boldsymbol{\lambda}, \boldsymbol{\pi})$ is differentiable with respect to $(\boldsymbol{\lambda}, \boldsymbol{\pi}) \in U$ and the partial differential coefficients are given by*

$$\frac{\partial w}{\partial \lambda_i} = \sum_{j=1}^{p} f_{ij}(\boldsymbol{x}_j) - {\mu_i}^{-1}(x_{p+1}),\ i = 1, \ldots, k, \tag{10.32}$$

$$\frac{\partial w}{\partial \pi_j} = \sum_{j=1}^{p} g_{jj}(\boldsymbol{x}_j),\ j = 1, \ldots, m. \tag{10.33}$$

Proof. Let $S_{n+1} = S \times [0,1] \in R^{n+1}$, where the notation $\times$ means the direct product. Then the set S_{n+1} is compact and convex from Assumption 10.2.1, and the objective function and the first and second constraints in the augmented primal problem (AP) are continuous on S_{n+1} from Assumption 10.2.2. Moreover, from Assumptions 10.2.2 and 10.2.3, the objective function is strictly convex on S_{n+1} and the constraints are also convex on S_{n+1}. Then, from Corollary 2 of Lasdon [58] (p. 427), the dual function $w(\boldsymbol{\lambda}, \boldsymbol{\pi})$ is differentiable for any $(\boldsymbol{\lambda}, \boldsymbol{\pi}) \in U$, and the partial derivatives of $w(\cdot)$ are given by (10.32) and (10.2.2), respectively.

Theorem 10.2.3. *The optimal solution of the augmented primal problem (AP) coincides with the optimal solution of the dual problem (AD).*

Proof. Let $(\boldsymbol{\lambda}^o, \boldsymbol{\pi}^o) \in U$ be an optimal solution of the augmented dual problem (AD) and $(\boldsymbol{x}^o, x_{p+1}^o)$ be the corresponding decision variables of the dual function, i.e.,

$$(\boldsymbol{x}^o, x_{p+1}^o) \in \{\boldsymbol{x} \in S, x_{p+1} \in [0,1] \mid (\boldsymbol{x}, x_{p+1}) \text{ minimizes } L(\boldsymbol{x}, x_{p+1}, \boldsymbol{\lambda}^o, \boldsymbol{\pi}^o)\}.$$

Since, from Theorem 10.2.1, $U = \{(\boldsymbol{\lambda}, \boldsymbol{\pi}) \in R^{k+m_0} \mid \boldsymbol{\lambda} \geq \mathbf{0},\ \boldsymbol{\pi} \geq \mathbf{0}\}$, and from Theorem 10.2.2, the dual function $w(\boldsymbol{\lambda}, \boldsymbol{\pi})$ is differentiable at $(\boldsymbol{\lambda}^o, \boldsymbol{\pi}^o)$, the following optimality conditions hold:

$$\left.\frac{\partial w(\boldsymbol{\lambda}, \boldsymbol{\pi})}{\partial \lambda_i}\right|_{(\boldsymbol{\lambda}, \boldsymbol{\pi})=(\boldsymbol{\lambda}^o, \boldsymbol{\pi}^o)} = \sum_{j=1}^{p} f_{ij}(\boldsymbol{x}_j^o) - {\mu_i}^{-1}(x_{p+1}^o) \quad \begin{cases} = 0 \ (\lambda_i^o > 0) \\ \leq 0 \ (\lambda_i^o = 0) \end{cases}$$

$$\left.\frac{\partial w(\boldsymbol{\lambda}, \boldsymbol{\pi})}{\partial \pi_j}\right|_{(\boldsymbol{\lambda}, \boldsymbol{\pi})=(\boldsymbol{\lambda}^o, \boldsymbol{\pi}^o)} = \sum_{j=1}^{p} g_{\ell j}(\boldsymbol{x}_j^o) \quad \begin{cases} = 0 \ (\pi_i^o > 0) \\ \leq 0 \ (\pi_i^o = 0) \end{cases}$$

These conditions imply that $(\boldsymbol{x}^o, x_{p+1}^o, \boldsymbol{\lambda}^o, \boldsymbol{\pi}^o)$ is a saddle point for $L(\cdot)$.

From these three theorems, it can be understood that the satisficing solution for the DM to the problem (10.18) can be obtained by solving the augmented dual problem (AD) for a sufficiently small positive ρ.

It should be noted here that, from (10.25), (10.26) and (10.27), the Lagrangian function $L(\cdot)$ is decomposable with respect to $(\boldsymbol{x}_1, \ldots, \boldsymbol{x}_p, x_{p+1})$ for some fixed Lagrange multipliers $(\boldsymbol{\lambda}, \boldsymbol{\pi})$. As a result, the dual function (10.29) can be decomposed into the following $(p+1)$ subproblems for some fixed Lagrange multipliers $(\boldsymbol{\lambda}, \boldsymbol{\pi})$.

Subproblem $P_j(\boldsymbol{\lambda}, \boldsymbol{\pi}) \quad (\, j = 1, \ldots, p\,)$

$$w_j(\boldsymbol{\lambda}, \boldsymbol{\pi}) \triangleq \min_{\boldsymbol{x}_j \in S_j} \ L_j(\boldsymbol{x}_j, \boldsymbol{\lambda}, \boldsymbol{\pi}) = \left\{ \sum_{i=1}^{k} \lambda_i f_{ij}(\boldsymbol{x}_j) + \sum_{\ell=1}^{m_0} \pi_j g_{\ell j}(\boldsymbol{x}_j) \right\} \tag{10.34}$$

Subproblem $P_{p+1}(\boldsymbol{\lambda})$

$$w_{p+1}(\boldsymbol{\lambda}) \triangleq \min_{x_{p+1} \in [0,1]} \ L_{p+1}(x_{p+1}, \boldsymbol{\lambda}) = \left\{ -x_{p+1} + \rho x_{p+1}^2 - \sum_{i=1}^{k} \lambda_i {\mu_i}^{-1}(x_{p+1}) \right\} \tag{10.35}$$

Then, from the definition of the dual function (10.29), it holds that

$$w(\boldsymbol{\lambda}, \boldsymbol{\pi}) = \sum_{j=1}^{p} w_j(\boldsymbol{\lambda}, \boldsymbol{\pi}) + w_{p+1}(\boldsymbol{\lambda}). \tag{10.36}$$

In order to improve the value of the dual function $w(\boldsymbol{\lambda}, \boldsymbol{\pi})$, we can apply a simple steepest decent method by making use of the partial differential coefficients (10.32) and (10.33) in Theorem 10.2.2.

Following the preceding discussions, we can now present the dual decomposition algorithm for deriving the satisficing solution for the DM to the problem (10.18) under the fuzzy decision [3].

Dual decomposition method with fuzzy goals

Step 1: Elicit a membership function from the DM for each of the objective functions of the problem (10.18) in a subjective manner.

Step 2: Set the iteration index $t = 0$, and set the initial values of the Lagrange multipliers $(\boldsymbol{\lambda}^t, \boldsymbol{\pi}^t) \geq \mathbf{0}$ appropriately.

Step 3: Solve the subproblems $P_j(\boldsymbol{\lambda}^t, \boldsymbol{\pi}^t)$, $j = 1, \ldots, p$, and $P_{p+1}(\boldsymbol{\lambda}^t)$ and obtain an optimal solution $\boldsymbol{x}_j(\boldsymbol{\lambda}^t, \boldsymbol{\pi}^t)$, $j = 1, \ldots, p$, $\boldsymbol{x}_{p+1}(\boldsymbol{\lambda}^t)$ and the corresponding optimal objective function values $w_j(\boldsymbol{\lambda}^t, \boldsymbol{\pi}^t)$, $j = 1, \ldots, p$, $w_{p+1}(\boldsymbol{\lambda}^t)$. In the following, for notational convenience, denote $\boldsymbol{x}^t = (\boldsymbol{x}_1(\boldsymbol{\lambda}^t, \boldsymbol{\pi}^t), \ldots, \boldsymbol{x}_p(\boldsymbol{\lambda}^t, \boldsymbol{\pi}^t))$ and $\boldsymbol{x}^t_{p+1} = \boldsymbol{x}_{p+1}(\boldsymbol{\lambda}^t)$, respectively.

Step 4: Obtain the dual function value $w(\boldsymbol{\lambda}^t, \boldsymbol{\pi}^t)$ as the summation of the optimal objective values $w_j(\boldsymbol{\lambda}^t, \boldsymbol{\pi}^t)$, $j = 1, \ldots, p$ and $w_{p+1}(\boldsymbol{\lambda}^t)$ (see (10.36)), and compute the direction vectors $\boldsymbol{D}_1(\boldsymbol{\lambda}^t) = (d_1(\lambda^t_1), \ldots, d_1(\lambda^t_k))$ and $\boldsymbol{D}_2(\boldsymbol{\pi}^t) = (d_2(\pi^t_1), \ldots, d_2(\pi^t_k))$, which improve the dual function value by the following formulae:

$$d_1(\lambda^t_i) = \begin{cases} \dfrac{\partial w}{\partial \lambda^t_i} = \displaystyle\sum_{j=1}^{p} f_{ij}(\boldsymbol{x}^t_j) - {\mu_i}^{-1}(\boldsymbol{x}^t_{p+1}) & ; \ \lambda^t_i > 0, \\ \max\left\{0, \ \dfrac{\partial w}{\partial \lambda^t_i}\right\} & ; \ \lambda^t_i = 0, \end{cases}$$

$$d_2(\pi^t_j) = \begin{cases} \dfrac{\partial w}{\partial \pi^t_j} = \displaystyle\sum_{j=1}^{p} g_{\ell j}(\boldsymbol{x}^t_j) & ; \ \pi^t_i > 0, \\ \max\left\{0, \ \dfrac{\partial w}{\partial \pi^t_j}\right\} & ; \ \pi^t_j = 0, \end{cases}$$

where $i = 1, \ldots, k$ and $j = 1, \ldots, m$.

Step 5: For the search direction vectors $\boldsymbol{D}_1(\boldsymbol{\lambda}^t) = (d_1(\lambda^t_1), \ldots, d_1(\lambda^t_k))$ and $\boldsymbol{D}_2(\boldsymbol{\pi}^t) = (d_2(\pi^t_1), \ldots, d_2(\pi^t_k))$ calculated in Step 4, solve the following one-dimensional search problem for obtaining the optimal step size α^t,

$$\begin{aligned} &\underset{\alpha \geq 0}{\text{maximize}} \ \ w(\boldsymbol{\lambda}^t + \alpha \boldsymbol{D}_1(\boldsymbol{\lambda}^t), \boldsymbol{\pi}^t + \alpha \boldsymbol{D}_2(\boldsymbol{\pi}^t)) \\ &\text{subject to} \ \ \boldsymbol{\lambda}^t + \alpha \boldsymbol{D}_1(\boldsymbol{\lambda}^t) \geq \mathbf{0}, \ \boldsymbol{\pi}^t + \alpha \boldsymbol{D}_2(\boldsymbol{\pi}^t) \geq \mathbf{0}. \end{aligned}$$

Step 6: If $\alpha^t \approx 0$, then stop. Otherwise, set $\boldsymbol{\lambda}^{t+1} = \boldsymbol{\lambda}^t + \alpha^t \boldsymbol{D}_1(\boldsymbol{\lambda}^t)$, $\boldsymbol{\pi}^{t+1} = \boldsymbol{\pi}^t + \alpha^t \boldsymbol{D}_2(\boldsymbol{\pi}^t)$, $t = t + 1$, and return to Step 3.

Example 10.2.1. In order to demonstrate the feasibility of the proposed algorithm, consider the following simple two objective nonlinear programming problem with the block angular structure.

$$\begin{array}{rl} \text{minimize} & f_1(\boldsymbol{x}) = f_{11}(x_1) + f_{12}(x_2) = (x_1 - 5)^2 + (x_2 - 3)^2 \\ \text{minimize} & f_2(\boldsymbol{x}) = f_{21}(x_1) + f_{22}(x_2) = (x_1 - 2)^2 + (x_2 - 4)^2 \\ \text{subject to} & g_1(\boldsymbol{x}) = g_{11}(x_1) + g_{12}(x_2) = (1/7)x_1 + (1/12)x_2 - 1 \leq 0 \\ & g_2(\boldsymbol{x}) = g_{21}(x_1) + g_{22}(x_2) = (1/18)x_1 + (1/6)x_2 - 1 \leq 0 \\ & 0 \leq x_1 \leq 7, \quad 0 \leq x_2 \leq 6. \end{array}$$

For this numerical example, suppose the hypothetical DM establishes the linear membership functions $\mu_1(f_1) = 1 - (1/6)f_1$ and $\mu_2(f_2) = 1 - (1/8)f_2$ for the two objectives $f_1(x)$ and $f_2(x)$, respectively. Then the augmented primal problem (AP) is formulated as

$$\begin{array}{rll} \text{minimize} & -x_3 + \rho x_3^2 & \\ \text{subject to} & (x_1 - 5)^2 + (x_2 - 3)^2 - 6(1 - x_3) & \leq 0 \\ & (x_1 - 2)^2 + (x_2 - 4)^2 - 8(1 - x_3) & \leq 0 \\ & (1/7)x_1 + (1/12)x_2 - 1 & \leq 0 \\ & (1/18)x_1 + (1/6)x_2 - 1 & \leq 0 \\ & 0 \leq x_1 \leq 7, \quad 0 \leq x_2 \leq 6, \quad 0 \leq x_3 \leq 1 & \end{array}$$

where $\rho = 0.01$.

Since this is a simple nonlinear programming problem, we first solve it directly by applying the well-known nonlinear programming code, called the revised version of the generalized reduced gradient program GRG2 [60]. As a result we obtain optimal decision values $x_1 = 3.6077$, $x_2 = 3.4641$, and the corresponding objective value $-x_3 + 0.01x_3^2 = 6.3692$, respectively. Moreover, the corresponding Lagrange multipliers to the constraints with respect to the objective functions become $\lambda_1 = 0.076367$, $\lambda_2 = 0.066122$, respectively.

With this result in mind, we now apply the proposed dual decomposition algorithm for solving it. By decomposing it into three subproblems, the corresponding subproblems $P_j(\cdot)$, $j = 1, 2, 3$, become

Subproblem $P_1(\lambda_1, \lambda_2, \pi_1, \pi_2)$:

$$\min_{0 \leq x_1 \leq 7} \left\{ \lambda_1(x_1 - 5)^2 + \lambda_2(x_1 - 2)^2 + \pi_1(x_1/7 - 1/2) + \pi_2(x_1/18 - 1/2) \right\}$$

Subproblem $P_2(\lambda_1, \lambda_2, \pi_1, \pi_2)$:

$$\min_{0 \leq x_2 \leq 6} \left\{ \lambda_1(x_2 - 3)^2 + \lambda_2(x_2 - 4)^2 + \pi_1(x_2/12 - 1/2) + \pi_2(x_2/6 - 1/2) \right\}$$

Subproblem $P_3(\lambda_1, \lambda_2)$:

$$\min_{0 \leq x_3 \leq 1} \left\{ -x_3 + 0.01x_3^2 - 6\lambda_1(1 - x_3) - 8\lambda_2(1 - x_3) \right\}$$

For these subproblems, applying the developed computer program yields the following result summarized in Table 10.3. From Table 10.3, we can see that the dual function value $w(\cdot)$ at the 30th iteration coincides with the calculation result through the revised version of GRG2 [60].

Table 10.3. Summary of the iteration

Iteration	λ_0	λ_1	x_1	x_2	$w(\cdot)$
0	0.1	0.1	3.5	3.5	−0.9
1	0.072	0.056	3.6875	3.4375	−0.67500
2	0.075828	0.062328	3.6466	3.4511	−0.64791
3	0.077863	0.065341	3.6312	3.4563	−0.63718
4	0.077641	0.065089	3.6319	3.4560	−0.63701
5	0.077669	0.065173	3.6312	3.4563	−0.63700
10	0.077448	0.065305	3.6276	3.4575	−0.63697
20	0.077064	0.065591	3.6206	3.4598	−0.63693
30	0.076815	0.065761	3.6163	3.4612	−0.63692

10.2.3 Primal decomposition method with fuzzy goals

To develop a primal decomposition method with fuzzy goals, in this subsection, we make the following assumptions, which are slightly stronger compared with the previous subsection:

Assumption 10.2.4. Each element of the vector-valued functions $\boldsymbol{h}_j(\boldsymbol{x}_j)$, $j = 1, \dots, p$, is convex and differentiable with respect to $\boldsymbol{x}_j$, and

$$S_j = \{\boldsymbol{x}_j \in R^{n_j} \mid \boldsymbol{h}_j(\boldsymbol{x}_j) \leq \boldsymbol{0}\} \neq \phi, \ j = 1, \dots, p \tag{10.37}$$

is convex and compact.

Observe that the set $S = \prod_{j=1,\dots,p} S_j \subset R^n$ becomes compact and convex.

Assumption 10.2.5. $f_{ij}(\boldsymbol{x}_j)$, $i = 1, \dots, k$, and $g_{\ell j}(\boldsymbol{x}_j)$, $\ell = 1, \dots, m_0$, are convex and differentiable on the subset S_j, $j = 1, \dots, p$.

Considering the vague nature of human's subjective judgments, it is quite natural to assume that the decision maker (DM) may have a fuzzy goal for each of the objective functions [78]. These fuzzy goals can be quantified by eliciting the corresponding membership functions $\mu_i(f_i(\boldsymbol{x}))$, $i = 1, \ldots, k$, through the interaction with the DM. Concerning the membership functions $\mu_i(f_i(\boldsymbol{x}))$, $i = 1, \ldots, k$, we make the following assumption.

Assumption 10.2.6. The membership function $\mu_i(\cdot)$ for each of the objective functions is strictly monotone decreasing and linear function on the subset S and defined as

$$\mu_i(f_i(\boldsymbol{x})) = \begin{cases} 1, & f_i(\boldsymbol{x}) \leq f_i^1 \\ \dfrac{f_i(\boldsymbol{x}) - f_i^0}{f_i^1 - f_i^0}, & f_i^1 \leq f_i(\boldsymbol{x}) \leq f_i^0 \\ 0, & f_i(\boldsymbol{x}) \geq f_i^0 \end{cases} \tag{10.38}$$

where f_i^0 and f_i^1 mean the maximum value of the unacceptable level and the minimum value of the acceptable level of the DM for the objective function $f_i(\boldsymbol{x})$ respectively. These values are subjectively assessed by the DM, which have to satisfy the following inequalities for any $\boldsymbol{x} \in S$.

$$f_i^1 \leq f_i(\boldsymbol{x}) \leq f_i^0, \quad i = 1, \ldots, k, \quad \forall \boldsymbol{x} \in S \tag{10.39}$$

After determining the membership functions, if we adopt the add-operator [131] for combining them, the satisficing solution for the DM can be obtained by solving the following nonlinear programming problem.

$$\left.\begin{array}{ll} \underset{\boldsymbol{x}}{\text{maximize}} & \left\{ \sum_{i=1}^{k} \mu_i(f_i(\boldsymbol{x})) \right\} \\ \text{subject to} & g_1(\boldsymbol{x}) = g_{11}(\boldsymbol{x}_1) + g_{12}(\boldsymbol{x}_2) + \cdots + g_{1p}(\boldsymbol{x}_p) \leq \boldsymbol{b}_1 \\ & g_2(\boldsymbol{x}) = g_{21}(\boldsymbol{x}_1) + g_{22}(\boldsymbol{x}_2) + \cdots + g_{2p}(\boldsymbol{x}_p) \leq \boldsymbol{b}_2 \\ & \vdots \\ & g_{m_0}(\boldsymbol{x}) = g_{m1}(\boldsymbol{x}_1) + g_{m2}(\boldsymbol{x}_2) + \cdots + g_{mp}(\boldsymbol{x}_p) \leq \boldsymbol{b}_{m_0} \\ & \boldsymbol{x}_j \in S_j, \quad j = 1, \ldots, p \end{array}\right\} \tag{10.40}$$

Here, from the definition of the membership functions, the objective function $\sum_{i=1}^{k} \mu_i(f_i(x))$ can be expressed as

$$\sum_{i=1}^{k} \mu_i(f_i(\boldsymbol{x})) = -\sum_{j=1}^{p} \left\{ \sum_{i=1}^{k} W_i f_{ij}(\boldsymbol{x}_j) \right\} + C. \tag{10.41}$$

where

$$W_i = \frac{1}{f_i^0 - f_i^1} > 0, \qquad C = \sum_{i=1}^{k} \frac{f_i^0}{f_i^0 - f_i^1} \tag{10.42}$$

Therefore, the problem (10.40) can equivalently expressed as

$$\left.\begin{array}{l} \underset{\boldsymbol{x}}{\text{minimize}} \ \displaystyle\sum_{j=1}^{p}\left\{\sum_{i=1}^{k} W_i f_{ij}(\boldsymbol{x}_j)\right\} \\ \text{subject to} \ g_1(\boldsymbol{x}) = g_{11}(\boldsymbol{x}_1) + g_{12}(\boldsymbol{x}_2) + \cdots + g_{1p}(\boldsymbol{x}_p) \leq \boldsymbol{b}_1 \\ \qquad\qquad g_2(\boldsymbol{x}) = g_{21}(\boldsymbol{x}_1) + g_{22}(\boldsymbol{x}_2) + \cdots + g_{2p}(\boldsymbol{x}_p) \leq \boldsymbol{b}_2 \\ \qquad\qquad \vdots \qquad\quad \vdots \qquad\qquad \vdots \qquad\qquad\qquad \vdots \qquad\quad \vdots \\ \qquad\qquad g_{m_0}(\boldsymbol{x}) = g_{m1}(\boldsymbol{x}_1) + g_{m2}(\boldsymbol{x}_2) + \cdots + g_{mp}(\boldsymbol{x}_p) \leq \boldsymbol{b}_{m_0} \\ \qquad\qquad\qquad \boldsymbol{x}_j \in S_j, \ \ j = 1, \ldots, p. \end{array}\right\} \tag{10.43}$$

It should be noted here that $\sum_{i=1}^{k} W_i f_{ij}(\boldsymbol{x}_j)$ is convex and differentiable on the set S_j from $W_i > 0$, $i = 1, \ldots, k$ and Assumption 10.2.5.

For this problem, if the number of the coupling constraints m is much smaller than the total number of constraints $\boldsymbol{h}_j(\boldsymbol{x}) \leq \boldsymbol{0}$, $j = 1, \ldots, p$, the decomposition method by right-hand-side allocation [26, 58] is especially efficient from the computational viewpoint.

Now, denote the vectors of the coupling constraint functions and the corresponding allocation vectors as follows:

$$\boldsymbol{g}_j(\boldsymbol{x}_j) = (g_{1j}(\boldsymbol{x}_j), \ldots, g_{m_0 j}(\boldsymbol{x}_j))^T, \quad j = 1, \ldots, p \tag{10.44}$$

$$\boldsymbol{y}_j = (y_{1j}, \ldots, y_{m_0 j})^T \in R^{m_0}, \quad j = 1, \ldots, p. \tag{10.45}$$

Then the allocation vectors $\boldsymbol{y}_j$, $j = 1, \ldots, p$ must satisfy the feasibility condition

$$\sum_{j=1}^{p} y_{\ell j} \leq b_j, \quad \ell = 1, \ldots, m_0. \tag{10.46}$$

Given the allocation vectors $\boldsymbol{y}_j$, $j = 1, \ldots, p$ satisfying the above conditions, we can decompose the problem (10.43) into the following p subproblems P_j, $j = 1, \ldots, p$.

$$\left.\begin{array}{l} \text{minimize} \ \displaystyle\sum_{i=1}^{k} W_i f_{ij}(\boldsymbol{x}_j) \\ \text{subject to} \ \boldsymbol{g}_j(\boldsymbol{x}_j) \leq \boldsymbol{y}_j, \quad \boldsymbol{x}_j \in S_j \end{array}\right\} \tag{10.47}$$

Since all of the above subproblems must be feasible, it is necessary to satisfy the following condition for the allocation vectors $\boldsymbol{y}_j$, $j = 1, \ldots, p$:

$$\boldsymbol{y}_j \in Y_j \triangleq \{\boldsymbol{y}_j \in R^{m_0} \mid \text{there exists } \boldsymbol{x}_j \in S_j \text{ such that } \boldsymbol{g}_j(\boldsymbol{x}_j) \leq \boldsymbol{y}_j\} \tag{10.48}$$

Also define the minimal subproblem objective values and their sum as follows.

$$w_j(\boldsymbol{y}_j) \triangleq \min \left\{ \sum_{i=1}^{k} W_i f_{ij}(\boldsymbol{x}_j) \,\middle|\, \boldsymbol{g}_j(\boldsymbol{x}_j) \leq \boldsymbol{y}_j,\ \boldsymbol{x}_j \in S_j \right\} \tag{10.49}$$

$$w(\boldsymbol{y}) \triangleq \sum_{j=1}^{p} w_j(\boldsymbol{y}_j) \tag{10.50}$$

Then the problem (10.43) can be equivalently transformed into the following form, where the allocation vectors $\boldsymbol{y}_j$, $j = 1, \ldots, p$, are the decision variables.

$$\left.\begin{array}{ll} \text{minimize} & w(\boldsymbol{y}) = \sum_{j=1}^{p} w_j(\boldsymbol{y}_j) \\ \text{subject to} & \sum_{j=1}^{p} y_{\ell j} \leq b_j, \quad \ell = 1, \ldots, m_0 \\ & \boldsymbol{y}_j \in Y_j, \quad j = 1, \ldots, p \end{array}\right\} \tag{10.51}$$

where $\boldsymbol{y}$ is the $(m_0 \times p)$-matrix defined as

$$\boldsymbol{y} = (\boldsymbol{y}_1, \ldots, \boldsymbol{y}_p) = \begin{pmatrix} y_{11} & y_{12} & \cdots & y_{1p} \\ y_{21} & y_{22} & \cdots & y_{2p} \\ \cdots & \cdots & \cdots & \cdots \\ y_{m_0 1} & y_{m_0 2} & \cdots & y_{m_0 p} \end{pmatrix}. \tag{10.52}$$

It should be noted here that the transformed problem (10.51) becomes the convex programming problem with respect to $\boldsymbol{y}$. From the properties of convexity and differentiability (Assumptions 10.2.4, 10.2.5 and 10.2.6), as a locally best usable direction, we can adopt a feasible direction which minimizes the directional derivative [58] of the transformed problem (10.51). Such a direction matrix can be obtained by solving the following direction-finding problem.

Direction-Finding Problem

$$\left.\begin{array}{ll} \underset{\boldsymbol{s}=(\boldsymbol{s}_1,\ldots,\boldsymbol{s}_p)}{\text{minimize}} & Dw(\boldsymbol{y}; \boldsymbol{s}) = \sum_{j=1}^{p} Dw_j(\boldsymbol{y}_j; \boldsymbol{s}_j) \\ \text{subject to} & \sum_{j=1}^{p} s_{\ell j} \leq 0, \ \ \ell \in B, \quad \|\boldsymbol{s}\| \leq 1 \end{array}\right\} \tag{10.53}$$

where $Dw(\boldsymbol{y};\boldsymbol{s})$ is a directional derivative of the objective function $w(\cdot)$ at $\boldsymbol{y}$ in the direction $\boldsymbol{s}$,

$$\boldsymbol{s} = (\boldsymbol{s}_1,\ldots,\boldsymbol{s}_p) = \begin{pmatrix} s_{11} & s_{12} & \cdots & s_{1p} \\ s_{21} & s_{22} & \cdots & s_{2p} \\ \cdots & \cdots & \cdots & \cdots \\ s_{m_01} & s_{m_02} & \cdots & s_{m_0p} \end{pmatrix} \tag{10.54}$$

is an $(m_0 \times p)$-search direction matrix at the allocation matrix $\boldsymbol{y}$,

$$B = \left\{ \ell \;\middle|\; b_\ell - \sum_{j=1}^{p} y_{\ell j} = 0 \right\} \tag{10.55}$$

is an index set of the binding constraints of (10.47) at $\boldsymbol{y}$, and $\|\boldsymbol{s}\|$ is a norm of $\boldsymbol{s}$.

Concerning the allocation matrix $\boldsymbol{y}$, we make the following assumptions.

Assumption 10.2.7. Any allocation vector $\boldsymbol{y}_j$ belongs to the interior of Y_j, i.e.,

$$\boldsymbol{y}_j \in \operatorname{int} Y_j, \quad j = 1,\ldots,p. \tag{10.56}$$

Assumption 10.2.8. The minimal subproblem objective values $w_j(\boldsymbol{y}_j)$, $j = 1,\ldots,p$, are differentiable with respect to $\boldsymbol{y}_j$.

Then, the direction-finding problem (10.53) can be reduced to the following simple linear programming problem [58]:

$$\left.\begin{aligned} &\underset{\boldsymbol{s}=(\boldsymbol{s}_1,\ldots,\boldsymbol{s}_p)}{\text{minimize}} && -\sum_{j=1}^{p} (\lambda_{1j},\ldots,\lambda_{m_0j}) \begin{pmatrix} s_{1j} \\ s_{2j} \\ \cdots \\ s_{m_0j} \end{pmatrix} \\ &\text{subject to} && \sum_{j=1}^{p} s_{\ell j} \le 0, \; \ell \in B \\ &&& -1 \le s_{\ell j} \le 1, \quad j = 1,\ldots,p\,;\; \ell = 1,\ldots,m_0 \end{aligned}\right\} \tag{10.57}$$

where $(\lambda_{1j},\ldots,\lambda_{m_0j})$ is the vector of the Lagrange multipliers of the constraint $\boldsymbol{g}_j(\boldsymbol{x}_j) \le \boldsymbol{y}_j$ in the subproblem (10.47).

Given a usable direction matrix $\boldsymbol{s}^t$, a new allocation matrix $\boldsymbol{y}^{t+1}$ is generated by choosing an appropriate step size $\alpha > 0$.

$$\boldsymbol{y}^{t+1} = \boldsymbol{y}^t + \alpha \boldsymbol{s}^t \tag{10.58}$$

where t denotes an iteration index.

To obtain such an optimal step size $\alpha > 0$, we solve the following convex one-dimensional search problem.

One-Dimensional Problem

$$\min_{\alpha \geq 0} \left\{ w(\boldsymbol{y}^t + \alpha \boldsymbol{s}^t) \;\middle|\; \sum_{j=1}^{p} (\boldsymbol{y}_j^t + \alpha \boldsymbol{s}_j^t) \leq \boldsymbol{b}_0,\ \ \boldsymbol{y}_j^t + \alpha \boldsymbol{s}_j^t \in Y_j,\ j = 1, \ldots, p \right\} \tag{10.59}$$

where $\boldsymbol{b}_0 = (b_1, \ldots, b_{m_0})^T$.

From the above discussions, we can now construct the primal decomposition method by right-hand-side allocation to the problem (10.18) for deriving the satisficing solution for the DM, where an add-operator [131] is adopted for combining the fuzzy goals of the DM.

Primal decomposition method with fuzzy goals

Step 1: Elicit a linear membership function $\mu_i(f_i(\boldsymbol{x}))$ satisfying the condition (10.39) from the DM for each of the objective functions $f_i(x)$, $i = 1, \ldots, k$, of the problem (10.18) in a subjective manner.

Step 2: Set the iteration index $t = 0$, and set the initial values of the allocation matrix

$$\boldsymbol{y}^t = (\boldsymbol{y}_1^t, \ldots, \boldsymbol{y}_p^t) = \begin{pmatrix} y_{11}^t & y_{12}^t & \cdots & y_{1p}^t \\ y_{21}^t & y_{22}^t & \cdots & y_{2p}^t \\ \cdots & \cdots & \cdots & \cdots \\ y_{m_01}^t & y_{m_02}^t & \cdots & y_{m_0p}^t \end{pmatrix} \tag{10.60}$$

satisfying the conditions (10.46) and (10.48) appropriately.

Step 3: Solve the p subproblems (10.47) and obtain an optimal solution $\boldsymbol{x}_j^t$, $j = 1, \ldots, p$, and the vector of the Lagrange multipliers $\boldsymbol{\lambda}_j^t = (\lambda_{1j}^t, \ldots, \lambda_{m_0j}^t)$, $j = 1, \ldots, p$, for the constraints $\boldsymbol{g}_j(\boldsymbol{x}_j) \leq \boldsymbol{y}_j$, respectively.

Step 4: Given the vector of the Lagrange multipliers $\boldsymbol{\lambda}_j^t$, $j = 1, \ldots, p$, in Step 3, solve the corresponding linear programming problem (10.57) to obtain the search direction matrix, and set the search direction matrix as

$$\boldsymbol{s}^t = (\boldsymbol{s}_1^t, \ldots, \boldsymbol{s}_p^t) = \begin{pmatrix} s_{11}^t & s_{12}^t & \cdots & s_{1p}^t \\ s_{21}^t & s_{22}^t & \cdots & s_{2p}^t \\ \cdots & \cdots & \cdots & \cdots \\ s_{m_01}^t & s_{m_02}^t & \cdots & s_{m_0p}^t \end{pmatrix} \tag{10.61}$$

Step 5: For the given search direction matrix s^t in Step 4, solve the following one-dimensional search problem to obtain the optimal step size $\alpha^t \geq 0$.

$$\min_{\alpha \geq 0}\left\{ w(y^t + \alpha s^t) \;\middle|\; \sum_{j=1}^{p}(y_j^t + \alpha s_j^t) \leq b_0,\; y_j^t + \alpha s_j^t \in Y_j,\; j = 1, \ldots, p \right\} \tag{10.62}$$

Step 6: If $\alpha^t \approx 0$, or $\|w(y^t) - w(y^{t-1})\| < \varepsilon$, then stop. Otherwise, set $y^{t+1} = y^t + \alpha^t s^t$, $t = t+1$, and return to Step 3.

Example 10.2.2. To demonstrate the feasibility of the proposed method, consider the following three-objective nonlinear programming problem with the block angular structure.

$$\begin{array}{rll}
\text{minimize } f_1(x) = & f_{11}(x_1, x_2) & + \quad f_{12}(x_3, x_4) \\
= & \left\{(x_1 - 8)^2 + (x_2 - 12)^2\right\} & + \left\{(x_3 - 30)^2 + (x_4 - 10)^2\right\} \\
\text{minimize } f_2(x) = & f_{21}(x_1, x_2) & + \quad f_{22}(x_3, x_4) \\
= & \left\{(x_1 - 10)^2 + (x_2 - 7)^2\right\} & + \left\{(x_3 - 8)^2 + (x_4 - 25)^2\right\} \\
\text{minimize } f_3(x) = & f_{31}(x_1, x_2) & + \quad f_{32}(x_3, x_4) \\
= & \left\{(x_1 - 35)^2 + (x_2 - 10)^2\right\} & + \left\{(x_3 - 12)^2 + (x_4 - 7)^2\right\} \\
\text{subject to } g_1(x) = & g_{11}(x_1, x_2) & + \quad g_{12}(x_3, x_4) \\
= & \{x_1/3 + x_2/10\} & + \quad \{x_3/7 + x_4/8\} \leq 1 \\
g_2(x) = & g_{21}(x_1, x_2) & + \quad g_{22}(x_3, x_4) \\
= & \{x_1/15 + x_2/12\} & + \quad \{x_3/5 + x_4/10\} \leq 1 \\
g_3(x) = & g_{31}(x_1, x_2) & + \quad g_{32}(x_3, x_4) \\
= & \{x_1/10 + x_2/12\} & + \quad \{x_3/8 + x_4/4\} \leq 1 \\
& x_j \geq 0, \quad j = 1, \ldots, 4 &
\end{array}$$

For this problem, suppose that the hypothetical DM establishes the following linear membership functions.

$$\mu_1(f_1(x)) = \frac{f_1^0 - f_1(x)}{f_1^0 - f_1^1} = \frac{1250 - f_1(x)}{1250 - 900}$$

$$\mu_2(f_2(x)) = \frac{f_2^0 - f_2(x)}{f_2^0 - f_2^1} = \frac{850 - f_2(x)}{850 - 650}$$

$$\mu_3(f_3(x)) = \frac{f_3^0 - f_3(x)}{f_3^0 - f_3^1} = \frac{1550 - f_3(x)}{1550 - 1300}$$

Then the problem to be solved for obtaining the satisficing solution for the DM can be formulated as

$$\begin{aligned}
\text{minimize} \quad & -\left\{ \frac{1250 - (f_{11}(x_1, x_2) + f_{12}(x_3, x_4))}{1250 - 900} \right. \\
& + \frac{850 - (f_{21}(x_1, x_2) + f_{22}(x_3, x_4))}{850 - 650} \\
& \left. + \frac{1550 - (f_{31}(x_1, x_2) + f_{32}(x_3, x_4))}{1550 - 1300} \right\} \\
\text{subject to} \quad & g_1(\boldsymbol{x}) = g_{11}(x_1, x_2) + g_{12}(x_3, x_4) \leq 1 \\
& g_2(\boldsymbol{x}) = g_{21}(x_1, x_2) + g_{22}(x_3, x_4) \leq 1 \\
& g_3(\boldsymbol{x}) = g_{31}(x_1, x_2) + g_{32}(x_3, x_4) \leq 1 \\
& x_j \geq 0, \quad j = 1, \ldots, 4
\end{aligned}$$

First applying the revised version of the generalized reduced gradient program called GRG2 [60] for solving the problem yields the optimal decision values $x_1^* = 0.449168$, $x_2^* = 1.82974$, $x_3^* = 3.31024$, $x_4^* = 1.55530$ and the corresponding objective value $w(\boldsymbol{y}^*) = -\sum_{i=1}^{3} \mu_i(f_i(\boldsymbol{x}^*)) = -2.41332$ respectively. Moreover, the corresponding optimal values of the allocation matrix become $g_{11}(x_1^*, x_2^*) = 0.332697$, $g_{12}(x_3^*, x_4^*) = 0.667304$, $g_{21}(x_1^*, x_2^*) = 0.182423$, $g_{22}(x_3^*, x_4^*) = 0.817578$, $g_{31}(x_1^*, x_2^*) = 0.197395$, $g_{32}(x_3^*, x_4^*) = 0.802605$, respectively.

Then to show the feasibility of the proposed algorithm, for the given initial values of the allocation matrix $y_{11}, y_{12}, y_{21}, y_{22}, y_{31}, y_{32}$ satisfying the condition (10.46) and (10.48), we formulate the two subproblems:

Subproblem $P_1(y_{11}, y_{21}, y_{31})$

$$\begin{aligned}
\underset{(x_1, x_2) \geq 0}{\text{minimize}} \; & -\left\{ \frac{1250 - f_{11}(x_1, x_2)}{1250 - 900} + \frac{850 - f_{21}(x_1, x_2)}{850 - 650} + \frac{1550 - f_{31}(x_1, x_2)}{1550 - 1300} \right\} \\
\text{subject to } & g_{11}(x_1, x_2) \leq y_{11} \\
& g_{21}(x_1, x_2) \leq y_{21} \\
& g_{31}(x_1, x_2) \leq y_{31}
\end{aligned}$$

Subproblem $P_2(y_{12}, y_{22}, y_{32})$

$$\begin{aligned}
\underset{(x_3, x_4) \geq 0}{\text{minimize}} \; & -\left\{ \frac{-f_{12}(x_3, x_4)}{1250 - 900} + \frac{-f_{22}(x_3, x_4)}{850 - 650} + \frac{-f_{32}(x_3, x_4)}{1550 - 1300} \right\} \\
\text{subject to } & g_{12}(x_3, x_4) \leq y_{12} \\
& g_{22}(x_3, x_4) \leq y_{22} \\
& g_{32}(x_3, x_4) \leq y_{32}
\end{aligned}$$

and apply the developed computer program for solving the example. The results obtained are summarized in Table 10.4.

Table 10.4. Summary of the iteration

Iteration	1	10	20	30
y_{11}	0.3	0.319922	0.314064	0.311050
y_{12}	0.6	0.680078	0.685936	0.688950
y_{21}	0.1	0.139066	0.150020	0.153120
y_{22}	0.8	0.829297	0.841038	0.843579
y_{31}	0.1	0.160547	0.167971	0.169868
y_{32}	0.8	0.839453	0.832029	0.830138
x_1	0.84375	0.604132	0.529106	0.502541
x_2	0.18750	1.185444	1.376956	1.435360
x_3	2.48889	3.239931	3.359016	3.408335
x_4	1.95556	1.737847	1.648609	1.616363
$w(\boldsymbol{y})$	−2.155384	−2.399757	−2.408099	−2.410648

In Table 10.4, the objective value $w(\boldsymbol{y})$ comes close to the optimal objective value $w(\boldsymbol{y}^*)$ rapidly until iteration 10. However, in the neighbourhood of the optimal value, the convergence becomes slow.

It is significant to point out here that the primal decomposition method with fuzzy goals presented in this section has already been extended to include nonlinear membership functions. Interested readers might refer to Yano and Sakawa [138] for details. However, a possible generalization of all the results presented in this section along the same line as in Chapter 6 would be required.

References

1. T. Bäck, *Evolutionary Algorithms in Theory and Practice*, Oxford University Press, 1996.
2. T. Bäck, D.B. Fogel and Z. Michalewicz, *Hand Book of Evolutionary Computation*, Oxford University Press, 1997.
3. R.E. Bellman and L.A. Zadeh, Decision making in a fuzzy environment, *Management Science*, Vol. 17, pp. 141–164, 1970.
4. R. Benayoun, J. de Montgofier, J. Tergny and O. Larichev, Linear programming with multiple objective functions, Step method (STEM), *Mathematical Programming*, Vol. 1, pp. 366–375, 1971.
5. M. Berkelaar, lp_solve 2.0, ftp://ftp.es.ele.tue.nl/ pub/lp_solve.
6. L. Chambers (ed.), *Practical Handbook of Genetic Algorithms: Applications*, Volume I, CRC Press, Boca Raton, 1995.
7. A. Charnes and W.W. Cooper, Programming with linear fractional functions, *Naval Research Logistic Quarterly*, Vol. 9, pp. 181–186, 1962.
8. V. Chankong and Y.Y. Haimes, *Multiobjective Decision Making: Theory and Methodology*, North-Holland, Amsterdam, 1983.
9. E.U. Choo and D.R. Atkins, An interactive algorithm for multicriteria programming, *Computers & Operations Research*, Vol. 7, pp. 81–87, 1980.
10. J.L. Cochrane and M. Zeleny, (eds.), *Multiple Criteria Decision Making*, University of South Carolina Press, Columbia, 1973.
11. R.J. Dakin, A tree search algorithm for mixed integer programming problems, *Computer Journal*, Vol. 8, pp. 250–255, 1965.
12. G.B. Dantzig, *Linear Programming and Extensions*, Princeton University Press, New Jersey, 1961.
13. G.B. Dantzig and P. Wolfe, Decomposition principle for linear programs, *Operations Research*, Vol. 8, pp. 101–111, 1960.
14. G.B. Dantzig and P. Wolfe, The decomposition algorithm for linear programming, *Econometrica*, Vol. 29, pp. 767–778, 1961.
15. D. Dasgupta and Z. Michalewicz (eds.), *Evolutionary Algorithms in Engineering Applications*, Springer, 1997.
16. L. Davis (ed.), *Genetic Algorithms and Simulated Annealing*, Morgan Kaufmann Publishers, 1987.
17. M. Delgado, J. Kacprzyk, J.-L. Verdegay and M.A. Vila (eds.), *Fuzzy Optimization: Recent Advances*, Physica-Verlag, Heidelberg, 1994.

18. D. Dubois and H. Prade, Operations on fuzzy numbers, *International Journal of Systems Science*, Vol. 9, pp. 613–626, 1978.
19. D. Dubois and H. Prade, *Fuzzy Sets and Systems: Theory and Applications*, Academic Press, New York, 1980.
20. D. Dubois and H. Prade, A review of fuzzy set aggregation connectives, *Information Sciences*, Vol. 36, pp. 85–121, 1985.
21. A.V. Fiacco, *Introduction to Sensitivity and Stability Analysis in Nonlinear Programming*, Academic Press, New York, 1983.
22. J. Fichefet, GPSTEM: an interactive multiobjective optimization method, *Progress in Operations Research*, North-Holland, Vol. 1, pp. 317–332, 1976.
23. M. Gen and R. Cheng, *Genetic Algorithms & Engineering Design*, Wiley-Interscience, New York, 1996.
24. A.M. Geoffrion, Elements of large-scale mathematical programming -PART I-, *Management Science*, Vol. 16, pp. 652–675, 1970.
25. A.M. Geoffrion, Elements of large-scale mathematical programming -PART II-, *Management Science*, Vol. 16, pp. 676–691, 1970.
26. A.M. Geoffrion, Primal resource-directive approaches for optimizing nonlinear decomposable systems, *Operations Research*, Vol. 18, pp. 375–403, 1970.
27. A.M. Geoffrion, J.S. Dyer, and A. Feinberg, An interactive approach for multicriterion optimization, with an application to the operation of an academic department, *Management Science*, Vol. 19, pp. 357–368, 1972.
28. D.E. Goldberg, *Genetic Algorithms in Search, Optimization, and Machine Learning*, Addison Wesley, 1989.
29. D.E. Goldberg and R. Lingle, Alleles, loci, and the traveling salesman problem, *Proceedings of the 1st International Conference on Genetic Algorithms and Their Applications*, Lawrence Erlbaum Associates, Publishers, New Jersey, 154–159, 1985.
30. J.J. Grefenstette, R. Gopal, B. Rosmaita and D. Van Gucht, Genetic algorithms for the traveling salesman problem, *Proceedings of the 1st International Conference on Genetic Algorithms and Their Applications*, Lawrence Erlbaum Associates, Publishers, New Jersey, 160–168, 1985.
31. Y.Y. Haimes and V. Chankong, Kuhn-Tucker multipliers as trade-offs in multiobjective decision-making analysis, *Automatica*, Vol. 15, pp. 59–72, 1979.
32. Y.Y. Haimes, K. Tarvainen, T. Shima and J. Thadathil, *Hierarchical Multiobjective Analysis of Large-Scale Systems*, Hemisphere Publishing Corporation, New York, 1989.
33. E.L. Hannan, Linear programming with multiple fuzzy goals, *Fuzzy Sets and Systems*, Vol. 6, pp. 235–248, 1981.
34. D.M. Himmelblau (ed.), *Decomposition of Large-Scale Problems*, North-Holland, Amsterdam, 1973.
35. J.K. Ho, Convergence behavior of decomposition algorithms for linear programs, *Operations Research Letters*, Vol. 3, pp. 91–94, 1984.
36. J.K. Ho, A parametric subproblems for dual methods in decomposition, *Mathematics of Operations Research*, Vol. 11, pp. 644–650, 1986.

37. J.K. Ho, Recent advances in the decomposition approach to linear programming, *Mathematical Programming Study*, Vol. 31, pp. 119–128, 1987.
38. J.K. Ho and E. Loute, An advanced implementation of the Dantzig-Wolfe decomposition algorithm for linear programming, *Mathematical Programming*, Vol. 20, pp. 303–326, 1981.
39. J.K. Ho and E. Loute, Computational experience with advanced implementation of decomposition algorithms for linear programming, *Mathematical Programming*, Vol. 27, pp. 283–290, 1983.
40. J.K. Ho and R.P. Sundarraj, An advanced implementation of the Dantzig-Wolfe decomposition algorithm for linear programming, *Mathematical Programming*, Vol. 20, pp. 303–326, 1981.
41. J.K. Ho and R.P. Sundarraj, Computational experience with advanced implementation of decomposition algorithms for linear programming, *Mathematical Programming*, Vol. 27, pp. 283–290, 1983.
42. J.K. Ho and R. P. Sundarraj, *DECOMP: an Implementation of Dantzig-Wolfe Decomposition for Linear Programming*, Springer-Verlag, 1989.
43. J.H. Holland, *Adaptation in Natural and Artificial Systems*, University of Michigan Press, 1975, MIT Press, 1992.
44. J. Kacprzyk and S.A. Orlovski (eds.), *Optimization Models Using Fuzzy Sets and Possibility Theory*, D. Reidel Publishing Company, Dordrecht, 1987.
45. N. Karmarkar, A new polynomial time algorithm for linear programming, *Combinatorics*, Vol. 4, pp. 373–395, 1984.
46. K. Kato and M. Sakawa, Genetic algorithms with decomposition procedures for fuzzy multiobjective 0-1 programming problems with block angular structure, *Proceedings of 1996 IEEE International Conference on Evolutionary Computation*, pp. 706–709, 1996.
47. K. Kato and M. Sakawa, An interactive fuzzy satisficing method for multiobjective structured 0-1 programs through genetic algorithms, *Proceedings of mini-Symposium on Genetic Algorithms and Engineering Design*, pp. 48–57, 1996.
48. K. Kato and M. Sakawa, An interactive fuzzy satisficing method for multiobjective linear fractional programs with block angular structure, *Cybernetics and Systems: An International Journal*, Vol. 28, No. 3, pp. 245–262, 1997.
49. K. Kato and M. Sakawa, Interactive decision making for multiobjective block angular 0-1 programming problems with fuzzy parameters through genetic algorithms, *Proceedings of the Sixth IEEE International Conference on Fuzzy Systems*, Vol. 3, pp. 1645–1650, 1997.
50. K. Kato and M. Sakawa, An interactive fuzzy satisficing method for multiobjective block angular 0-1 programming problems involving fuzzy parameters through genetic algorithms with decomposition procedures, *Proceedings of the Seventh International Fuzzy Systems Association World Congress*, Vol. 3, pp. 9–14, 1997.
51. K. Kato and M. Sakawa, An interactive fuzzy satisficing method for large-scale multiobjective 0-1 programming problems with fuzzy parameters through

genetic algorithms, *European Journal of Operational Research*, Vol. 107, No. 3, pp. 590–598, 1998.

52. K. Kato and M. Sakawa, Large scale fuzzy multiobjective 0-1 programs through genetic algorithms with decomposition procedures, *Proceedings of Second International Conference on Knowledge-Based Intelligent Electronic Systems*, Vol. 1, pp. 278–284, 1998.
53. K. Kato and M. Sakawa, Improvement of genetic algorithm by decomposition procedures for fuzzy block angular multiobjective knapsack problems, *Proceedings of the Eighth International Fuzzy Systems Association World Congress*, Vol. 1, pp. 349–353, 1999.
54. J.S.H. Kornbluth and R.E. Steuer, Multiple objective linear fractional programming, *Management Science*, Vol. 27, pp. 1024–1039, 1981.
55. J.S.H. Kornbluth and R.E. Steur, Goal programming with linear fractional criteria, *European Journal of Operation Research*, Vol. 8, pp. 58–65, 1981.
56. Y.J. Lai and C.L. Hwang, *Fuzzy Multiple Objective Decision Making: Methods and Applications*, Springer-Verlag, Berlin, 1994.
57. L.S. Lasdon, Duality and decomposition in mathematical programming, *IEEE Transaction on Systems Science and Cybernetics*, Vol. SSC-4, pp. 86–100, 1968.
58. L.S. Lasdon, *Optimization Theory for Large Scale Systems*, Macmillan, 1970.
59. L.S. Lasdon, R.L. Fox and M.W. Ratner, Nonlinear optimization using the generalized reduced gradient method, *Revue Française d'Automatique,Informatique et Researche Opérationnelle*, Vol. 3, pp. 73–103, 1974.
60. L.S. Lasdon, A.D. Waren and M.W. Ratner, *GRG2 User's Guide*, Technical Memorandum, University of Texas (1980).
61. H. Leberling, On finding compromise solution in multicriteria problems using the fuzzy min-operator, *Fuzzy Sets and Systems*, Vol. 6, pp. 105–228, 1981.
62. M.K. Luhandjula, Compensatory operators in fuzzy linear programming with multiple objectives, *Fuzzy Sets and Systems*, Vol. 8, pp. 245-252, 1982.
63. M.K. Luhandjula, Fuzzy approaches for multiple objective linear fractional optimization, *Fuzzy Sets and Systems*, Vol. 13, pp. 11-23, 1984.
64. M.K. Luhandjula, Fuzzy optimization: an appraisal, *Fuzzy Sets and Systems*, Vol. 30, pp. 257–282, 1989.
65. J.G. March and H.A. Simon, *Organizations*, Wiley, New York, 1958.
66. R.K. Martin, *Large Scale Linear and Integer Optimization: A Unified Approach*, Kluwer Academic Publishers, Boston, 1999.
67. Z. Michalewicz, *Genetic Algorithms + Data Structures = Evolution Programs*, Springer-Verlag, Berlin, 1992, Second extended edition, 1994, Third revised and extended edition, 1996.
68. K.R. Oppenheimer, A proxy approach to multiattribute decision making, *Management Science*, Vol. 24, pp. 675–689, 1978.
69. W. Orchard-Hays, *Advanced Linear Programming: Computing Techniques*, McGraw-Hill, New York, 1968.
70. S.A. Orlovski, Multiobjective programming problems with fuzzy parameters, *Control and Cybernetics*, Vol. 13, pp. 175–183, 1984.

71. W. Pedrycz (ed.), *Fuzzy Evolutionary Computation*, Kluwer Academic Publishers, Boston, 1997.
72. K. Ritter, A decomposition method for linear programming problems with coupling constraints and variables, *Mathematics Research Center, University of Wisconsin Rept. 739*, 1967.
73. H. Rommelfanger, Interactive decision making in fuzzy linear optimization problems, *European Journal of Operational Research*, Vol. 41, pp. 210–217, 1989.
74. H. Rommelfanger, Fuzzy linear programming and applications, *European Journal of Operational Research*, Vol. 92, pp. 512–527, 1989.
75. M. Sakawa, An interactive computer program for multiobjective decision making by the sequential proxy optimization technique, *International Journal of Man-Machine Studies*, Vol. 14, pp. 193–213, 1981.
76. M. Sakawa, Interactive computer programs for fuzzy linear programming with multiple objectives, *International Journal of Man-Machine Studies*, Vol. 18, pp. 489–503, 1983.
77. M. Sakawa, Interactive fuzzy decision making for multiobjective nonlinear programming problems, in M. Grauer and A. P. Wierzbicki (eds.) *Interactive Decision Analysis*, Springer-Verlag, Berlin, pp. 105–112, 1984.
78. M. Sakawa, *Fuzzy Sets and Interactive Multiobjective Optimization*, Plenum Press, New York, 1993.
79. M. Sakawa, M. Inuiguchi, K. Kato and K. Sawada, A fuzzy programming approach to structured linear programs, in Z. Bien and K. C. Min (eds.) *Fuzzy Logic and Its Applications to Engineering, Information Sciences, and Intelligent Systems*, Kluwer Academic Publishers, Dordrecht, pp. 455–464, 1995.
80. M. Sakawa, M. Inuiguchi and K. Sawada, Fuzzy programming approach to large-scale linear problems, *Proceedings of the Fifth International Fuzzy Systems Association World Congress*, Vol. 1, pp. 625–655, 1993.
81. M. Sakawa, M. Inuiguchi and K. Sawada, A fuzzy satisficing method for large-scale multiobjective linear programming problems with block angular structure, *Fuzzy Sets and Systems*, Vol. 78, No. 3, pp. 279–288, 1996.
82. M. Sakawa, M. Inuiguchi, H. Sunada and K. Sawada, Fuzzy multiobjective combinatorial optimization through revised genetic algorithms, *Japanese Journal of Fuzzy Theory and Systems*, Vol. 6, No. 1, pp. 177-186, 1994 (in Japanese).
83. M. Sakawa and K. Kato, Fuzzy programming for large-scale multiobjective linear programming problems with fuzzy parameters, *Large Scale Systems: Theory and Applications, Preprints of the 7th IFAC/IFORS/IMACS Symposium*, Vol. 1, pp. 451–456, 1995.
84. M. Sakawa and K. Kato, An interactive fuzzy satisficing method for large-scale multiobjective linear programs with fuzzy numbers, *Proceedings of the International Joint Conference of the Fourth IEEE International Conference on Fuzzy Systems and the Second International Fuzzy Engineering Symposium*, Vol. 3, pp. 1155–1162, 1995.

85. M. Sakawa and K. Kato, An interactive fuzzy satisficing method for multiobjective structured linear programs and its application, *1995 IEEE International Conference on Systems, Man and Cybernetics*, Vol. 5, pp. 4045–4050, 1995.
86. M. Sakawa and K. Kato, An interactive fuzzy satisficing method for large scale multiobjective linear programming problems with fuzzy parameters, *Journal of Japan Society for Fuzzy Theory and Systems*, Vol. 8, No. 3, pp. 547–557, 1996 (in Japanese).
87. M. Sakawa and K. Kato, Interactive decision making for large-scale multiobjective linear programs with fuzzy numbers, *Fuzzy Sets and Systems*, Vol. 88, No. 2, pp. 161–172, 1997.
88. M. Sakawa and K. Kato, An interactive fuzzy satisficing method for multiobjective block angular linear fractional programming problems with fuzzy parameters, *Proceedings of the Seventh International Fuzzy Systems Association World Congress*, Vol. 3, pp. 21–26, 1997.
89. M. Sakawa and K. Kato, Interactive decision-making for multiobjective linear fractional programming problems with block angular structure involving fuzzy numbers, *Fuzzy Sets and Systems*, ol. 97, No. 1, pp. 19–31, 1998.
90. M. Sakawa and K. Kato, An interactive fuzzy satisficing method for structured multiobjective linear fractional programs with fuzzy numbers, *European Journal of Operational Research*, Vol. 107, No. 3, pp. 575–589, 1998.
91. M. Sakawa and K. Kato, An interactive fuzzy satisficing method for multiobjective block angular linear programming problems with fuzzy parameters, *Fuzzy Sets and Systems*, Vol. 107, No. 1, pp. 55–69, 2000.
92. M. Sakawa, K. Kato and H. Mohara, Efficiency of a decomposition method for large-scale multiobjective fuzzy linear programming problems with block angular structure, *Proceedings of Second International Conference on Knowledge-Based Intelligent Electronic Systems*, Vol. 1, pp. 80–86, 1998.
93. M. Sakawa, K. Kato and H. Mohara, Efficiency of decomposition method for structured multiobjective linear programming problems with fuzzy numbers, *Cybernetics and Systems: An International Journal*, Vol. 30, No. 6, pp. 551–570, 1999.
94. M. Sakawa, K. Kato and R. Mizouchi, Interactive decision making for large scale multiobjective linear programming problems with fuzzy parameters, *Journal of Japan Society for Fuzzy Theory and Systems*, Vol. 7, No. 3, pp. 612–623, 1995 (in Japanese).
95. M. Sakawa, K. Kato, H. Obata and K. Ooura, An approximate solution method for general multiobjective 0-1 programming problems through genetic algorithms with double string representation, *Transactions of the Institute of Electronics, Information and Communication Engineers*, Vol. J82-A, No. 7, pp. 1066–1073, 1999 (in Japanese).
96. M. Sakawa, K. Kato and T. Shibano, An interactive fuzzy satisficing method for multiobjective multidimensional 0-1 knapsack problems through genetic algorithms, *Proceedings of 1996 IEEE International Conference on Evolutionary Computation*, pp. 243–246, 1996.

97. M. Sakawa, K. Kato, H. Sunada and Y. Enda, An interactive fuzzy satisficing method for multiobjective 0-1 programming problems through revised genetic algorithms, *Journal of Japan Society for Fuzzy Theory and Systems*, Vol. 7, No. 2, pp. 361-370, 1995 (in Japanese).
98. M. Sakawa, K. Kato, H. Sunada and T. Shibano, Fuzzy programming for multiobjective 0-1 programming problems through revised genetic algorithms, *European Journal of Operational Research*, Vol. 97, pp. 149–158, 1997.
99. M. Sakawa, K. Kato and S. Ushiro, An interactive fuzzy satisficing method for multiobjective 0-1 programming problems involving positive and negative coefficients through genetic algorithms with double strings, *Proceedings of the Eighth International Fuzzy Systems Association World Congress*, Vol. 1, pp. 430–434, 1999.
100. M. Sakawa, K. Kato, S. Ushiro and K. Ooura, Fuzzy programming for general multiobjective 0-1 programming problems through genetic algorithms with double strings, *1999 IEEE International Fuzzy Systems Conference Proceedings*, Vol. III, pp. 1522–1527, 1999.
101. M. Sakawa and K. Sawada, Fuzzy 0-1 programming through neural networks, in R. Lowen and M. Roubens (eds.), *Fuzzy Logic, State of Art*, Kluwer Academic Publishers, Dordrecht, pp. 311–320, 1993.
102. M. Sakawa and K. Sawada, An interactive fuzzy satisficing method for large-scale multiobjective linear programming problems with block angular structure, *Fuzzy Sets and Systems: Special Issue on Operations Research*, Vol. 67, pp. 5–17, 1994.
103. M. Sakawa, K. Sawada and M. Inuiguchi, A fuzzy satisficing method for large-scale linear programming problems with block angular structure, *European Journal of Operational Research*, Vol. 81, pp. 399–409, 1995.
104. M. Sakawa and F. Seo, Interactive multiobjective decisionmaking for large-scale systems and its application to environmental systems, *IEEE Transactions Systems, Man and Cybernetics*, Vol. SMC-10, pp. 796–806, 1980.
105. M. Sakawa and F. Seo, Interactive multiobjective decision making in environmental systems using sequential proxy optimization techniques (SPOT), *Automatica*, Vol. 18, pp. 155–165, 1982.
106. M. Sakawa and T. Shibano, Interactive fuzzy programming for multiobjective 0-1 programming problems through genetic algorithms with double strings, in Da Ruan (ed.) *Fuzzy Logic Foundations and Industrial Applications*, Kluwer Academic Publishers, Boston, pp. 111–128, 1996.
107. M. Sakawa and T. Shibano, Multiobjective fuzzy satisficing methods for 0-1 knapsack problems through genetic algorithms, in W. Pedrycz (ed.) *Fuzzy Evolutionary Computation*, Kluwer Academic Publishers, Boston, pp. 155-177, 1997.
108. M. Sakawa and T. Shibano, An interactive fuzzy satisficing method for multiobjective 0-1 programming problems with fuzzy numbers through genetic algorithms with double strings, *European Journal of Operational Research*, Vol. 107, pp. 564–574, 1998.

109. M. Sakawa and T. Shibano, An interactive approach to fuzzy multiobjective 0-1 programming problems using genetic algorithms, in M. Gen and Y. Tsujimura (eds.) *Evolutionary Computations and Intelligent Systems*, Gordon & Breach, Inc. (to appear).
110. M. Sakawa and M. Tanaka, *Genetic Algorithms*, Asakura Publishing, 1995 (in Japanese).
111. M. Sakawa and T. Yumine, Interactive fuzzy decision-making for multiobjective linear fractional programming problems, *Large Scale Systems*, Vol. 5, pp. 105–114, 1983.
112. M. Sakawa and H. Yano, An interactive fuzzy satisficing method using augmented minimax problems and its application to environmental systems, *IEEE Transactions Systems, Man and Cybernetics*, Vol. SMC-15, No. 6, pp. 720–729, 1985.
113. M. Sakawa and H. Yano, Interactive decision making for multiobjective linear fractional programming problems with fuzzy parameters, *Cybernetics and Systems: An International Journal*, Vol. 16, pp. 377–394, 1985.
114. M. Sakawa and H. Yano, Interactive decision making for multiobjective linear problems with fuzzy parameters, in G. Fandel, M. Grauer, A. Kurzhanski and A. P. Wierzbicki (eds.) *Large-Scale Modeling and Interactive Decision Analysis*, pp. 88–96, 1986.
115. M. Sakawa and H. Yano, An interactive fuzzy satisficing method for multiobjective linear programming problems with fuzzy parameters, *Large Scale Systems: Theory and Applications, Proceedings of the IFAC/IFORS Symposium*, 1986.
116. M. Sakawa and H. Yano, An interactive fuzzy satisficing method for multiobjective nonlinear programming problems with fuzzy parameters, in R. Trappl (ed.) *Cybernetics and Systems '86*, D. Reidel Publishing Company, pp. 607–614, 1986.
117. M. Sakawa and H. Yano, An interactive fuzzy satisficing method for multiobjective linear fractional programming problems, *Fuzzy Sets and Systems*, Vol. 28, pp. 129–144, 1988.
118. M. Sakawa and H. Yano, Interactive decision making for multiobjective nonlinear programming problems with fuzzy parameters, *Fuzzy Sets and Systems*, Vol. 29, pp. 129–144, 1989.
119. M. Sakawa and H. Yano, An interactive fuzzy satisficing method for multiobjective nonlinear programming problems with fuzzy parameters, *Fuzzy Sets and Systems*, Vol. 30, pp. 221–238, 1989.
120. M. Sakawa and H. Yano, Trade-off rates in the hyperplane method for multiobjective optimization problems, *European Journal of Operational Research*, Vol. 44, pp. 105–118, 1990.
121. M. Sakawa and H. Yano, A fuzzy dual decomposition method for large-scale multiobjective nonlinear programming problems, *Fuzzy Sets and Systems: Special Issue on Operations Research*, Vol. 67, pp. 19–27, 1995.

122. M. Sakawa, H. Yano and K. Sawada, Primal decomposition method for multiobjective structured nonlinear programs with fuzzy goals, *Cybernetics and Systems: An International Journal*, Vol. 26, pp. 413–426, 1995.
123. M. Sakawa, H. Yano and T. Yumine, An interactive fuzzy satisficing method for multiobjective linear-programming problems and its application, *IEEE Transactions Systems, Man and Cybernetics*, Vol. SMC-17, No. 4, pp. 654–661, 1987.
124. K. Sawada and M. Sakawa, Fuzzy programming for Large-scale multiobjective linear programming problems, in M. Fushimi and K. Tone (eds.), *Proceedings of APORS'94*, World Scientific, pp. 515–522, 1995.
125. S. Schaible and T. Ibaraki, Fractional programming, *European Journal of Operational Research*, Vol. 12, pp. 325–338, 1983.
126. H-P. Schwefel, *Evolution and Optimum Seeking*, John Wiley & Sons, New York, 1995.
127. F. Seo and M. Sakawa, *Multiple Criteria Decision Analysis in Regional Planning: Concepts, Methods and Applications*, D. Reidel Publishing Company, Dordrecht, 1988.
128. R. Slowinski and J. Teghem (eds.), *Stochastic versus Fuzzy Approaches to Multiobjective Mathematical Programming Problems under Uncertainty*, Kluwer Academic Publishers, Dordrecht, 1990.
129. R.E. Steuer, *Multiple Criteria Optimization: Theory, Computation, and Application*, John Wiley & Sons, New York, 1986.
130. R.E. Steuer and E.U. Choo, An interactive weighted Tchebycheff procedure for multiple objective programming, *Mathematical Programming*, Vol. 26, pp. 326–344, 1983.
131. G. Sommer and M.A. Pollatschek, A fuzzy programming approach to an air pollution regulation problem, in R. Trappl and G.J. Klir and L. Ricciardi (eds.), *Progress in Cybernetics and Systems Research*, Hemisphere, pp. 303–323, 1978.
132. M.J. Todd, A Dantzig-Wolfe-like variant of Karmarkar's interior-point linear programming algorithm, *Operations Research*, Vol. 38, pp. 1006–1018, 1990.
133. E.L. Ulungu and J. Teghem, Multi-objective combinatorial optimization problems: a survey, *Journal of Multicriteria Decision Analysis*, Vol. 3, pp. 83–104, 1994.
134. J.-L. Verdegay and M. Delgado (eds.), *The Interface between Artificial Intelligence and Operations Research in Fuzzy Environment*, Verlag TÜV Rheinland, Köln, 1989.
135. A.P. Wierzbicki, The use of reference objectives in multiobjective optimization, in G. Fandel and T. Gal (eds.) *Multiple Criteria Decision Making: Theory and Application*, Springer-Verlag, Berlin, pp. 468–486, 1980.
136. A.P. Wierzbicki, A mathematical basis for satisficing decision making, *Mathematical Modeling*, Vol. 3, pp. 391–405, 1982.
137. D.A. Wismer (ed.), *Optimization Methods for Large-Scale Systems ... with applications*, McGraw-Hill, New York, 1971.

138. H. Yano and M. Sakawa, A three-level optimization method for fuzzy large-scale multiobjective nonlinear programming problems, *Fuzzy Sets and Systems*, Vol. 81, pp. 141–155, 1996.
139. L.A. Zadeh, Fuzzy sets, *Information and Control*, Vol. 8, pp. 338–353, 1974.
140. M. Zeleny, *Multiple Criteria Decision Making*, McGraw-Hill, New York, 1982.
141. H.-J. Zimmermann, Description and optimization of fuzzy systems, *International Journal of General Systems*, Vol. 2, pp. 209–215, 1976.
142. H.-J. Zimmermann, Fuzzy programming and linear programming with several objective functions, *Fuzzy Sets and Systems*, Vol. 1, pp. 45–55, 1978.
143. H.-J. Zimmermann, Fuzzy mathematical programming, *Computers & Operations Research*, Vol. 10, pp. 291–298, 1983.
144. H.-J. Zimmermann, *Fuzzy Sets, Decision-Making and Expert Systems*, Kluwer Academic Publishers, Boston, 1987.
145. H.-J. Zimmermann, *Fuzzy Set Theory and Its Applications*, Kluwer Academic Publishers, Boston, 1985, Second Edition, 1991, Third edition, 1996.
146. H.-J. Zimmermann and P. Zysno, Latent connectives in human decision making, *Fuzzy Sets and Systems*, Vol. 4, pp. 37-51, 1980.

Index

Studies in Fuzziness and Soft Computing

Vol. 25. J. Buckley and Th. Feuring
Fuzzy and Neural: Interactions and Applications, 1999
ISBN 3-7908-1170-X

Vol. 26. A. Yazici and R. George
Fuzzy Database Modeling, 1999
ISBN 3-7908-1171-8

Vol. 27. M. Zaus
Crisp and Soft Computing with Hypercubical Calculus, 1999
ISBN 3-7908-1172-6

Vol. 28. R.A. Ribeiro, H.-J. Zimmermann, R.R. Yager and J. Kacprzyk (Eds.)
Soft Computing in Financial Engineering, 1999
ISBN 3-7908-1173-4

Vol. 29. H. Tanaka and P. Guo
Possibilistic Data Analysis for Operations Research, 1999
ISBN 3-7908-1183-1

Vol. 30. N. Kasabov and R. Kozma (Eds.)
Neuro-Fuzzy Techniques for Intelligent Information Systems, 1999
ISBN 3-7908-1187-4

Vol. 31. B. Kostek
Soft Computing in Acoustics, 1999
ISBN 3-7908-1190-4

Vol. 32. K. Hirota and T. Fukuda
Soft Computing in Mechatronics, 1999
ISBN 3-7908-1212-9

Vol. 33. L.A. Zadeh and J. Kacprzyk (Eds.)
Computing with Words in Information/Intelligent Systems 1, 1999
ISBN 3-7908-1217-X

Vol. 34. L.A. Zadeh and J. Kacprzyk (Eds.)
Computing with Words in Information/Intelligent Systems 2, 1999
ISBN 3-7908-1218-8

Vol. 35. K.T. Atanassov
Intuitionistic Fuzzy Sets, 1999
ISBN 3-7908-1228-5

Vol. 36. L.C. Jain (Ed.)
Innovative Teaching and Learning, 2000
ISBN 3-7908-1246-3

Vol. 37. R. Słowiński and M. Hapke (Eds.)
Scheduling Under Fuzziness, 2000
ISBN 3-7908-1249-8

Vol. 38. D. Ruan (Ed.)
Fuzzy Systems and Soft Computing in Nuclear Engineering, 2000
ISBN 3-7908-1251-X

Vol. 39. O. Pons, M.A. Vila and J. Kacprzyk (Eds.)
Knowledge Management in Fuzzy Databases, 2000
ISBN 3-7908-1255-2

Vol. 40. M. Grabisch, T. Murofushi and M. Sugeno (Eds.)
Fuzzy Measures and Integrals, 2000
ISBN 3-7908-1255-2

Vol. 41. P. Szczepaniak, P. Lisboa and J. Kacprzyk (Eds.)
Fuzzy Systems in Medicine, 2000
ISBN 3-7908-1263-4

Vol. 42. S. Pal, G. Ashish and M. Kundu (Eds.)
Soft Computing for Image Processing, 2000
ISBN 3-7908-1217-X

Vol 43. L.C. Jain, B. Lazzerini and U. Halici (Eds.)
Innovations in ART Neural Networks, 2000
ISBN 3-7908-1270-6

Vol. 44. J. Aracil, F. Gordillo (Eds.)
Stability Issues in Fuzzy Control
ISBN 3-7908-1277-3

Vol. 45. N. Kasabov (Ed.)
Future Directions for Intelligent Information Systems on Information Sciences
ISBN 3-7908-1276-5

Vol. 46. J.N. Mordeson, P.S. Nair
Fuzzy Graphs and Fuzzy Hypergraphs
ISBN 3-7908-1286-2

Vol. 47. Ernest Czogała†, Jacek Łęski
Fuzzy and Neuro-Fuzzy Intelligent Systems
ISBN 3-7908-1289-7

GPSR Compliance
The European Union's (EU) General Product Safety Regulation (GPSR) is a set of rules that requires consumer products to be safe and our obligations to ensure this.

If you have any concerns about our products, you can contact us on

ProductSafety@springernature.com

In case Publisher is established outside the EU, the EU authorized representative is:

Springer Nature Customer Service Center GmbH
Europaplatz 3
69115 Heidelberg, Germany

www.ingramcontent.com/pod-product-compliance
Lightning Source LLC
LaVergne TN
LVHW012329060326
832902LV00011B/1790
* 9 7 8 3 6 6 2 0 0 3 8 6 2 *